Smart Proxy Modeling

Numerical simulation models are used in all engineering disciplines for modeling physical phenomena to learn how the phenomena work, and to identify problems and optimize behavior. Smart Proxy Models provide an opportunity to replicate numerical simulations with very high accuracy and can be run on a laptop within a few minutes, thereby simplifying the use of complex numerical simulations, which can otherwise take tens of hours. This book focuses on Smart Proxy Modeling and provides readers with all the essential details on how to develop Smart Proxy Models using Artificial Intelligence and Machine Learning, as well as how they may be used in real-world cases.

- Covers replication of highly accurate numerical simulations using Artificial Intelligence and Machine Learning
- Details application in reservoir simulation and modeling and computational fluid dynamics
- Includes real case studies based on commercially available simulators

Smart Proxy Modeling is ideal for petroleum, chemical, environmental, and mechanical engineers, as well as statisticians and others working with applications of data-driven analytics.

Smart Proxy Modeling
Artificial Intelligence and Machine Learning in Numerical Simulation

Shahab D. Mohaghegh

CRC Press
Taylor & Francis Group
Boca Raton London New York

CRC Press is an imprint of the
Taylor & Francis Group, an **informa** business

First edition published 2023
by CRC Press
6000 Broken Sound Parkway NW, Suite 300, Boca Raton, FL 33487-2742

and by CRC Press
4 Park Square, Milton Park, Abingdon, Oxon, OX14 4RN

CRC Press is an imprint of Taylor & Francis Group, LLC

© 2023 Shahab D. Mohaghegh

Library of Congress Cataloging-in-Publication Data
Names: Mohaghegh, Shahab D., author.
Title: Smart proxy modeling : artificial intelligence and machine learning in numerical simulation / Shahab D. Mohaghegh.
Description: First edition. | Boca Raton, FL : CRC Press, 2023. | Includes bibliographical references and index.
Identifiers: LCCN 2022019499 | ISBN 9781032151144 (hbk) | ISBN 9781032151151 (pbk) | ISBN 9781003242581 (ebk)
Subjects: LCSH: Engineering models--Data processing. | Engineering--Data processing. | Computer simulation. | Mathematical models. | Artificial intelligence. | Machine leairning. | Numerical analysis.
Classification: LCC TA342 .M643 2023 | DDC 620.001/1--dc23/eng/20220818
LC record available at https://lccn.loc.gov/2022019499

ISBN: 978-1-032-15114-4 (hbk)
ISBN: 978-1-032-15115-1 (pbk)
ISBN: 978-1-003-24258-1 (ebk)

DOI: 10.1201/9781003242581

Typeset in Sabon
by KnowledgeWorks Global Ltd.

This book is dedicated to Luna Stüttgen, my current most lovely and beautiful little girl, and his wonderful parents, my niece Ghazal Tajdini and Peter Stüttgen.

Contents

Preface

During the 21st century, Artificial Intelligence and Machine Learning have become the most significant science and technology that would create a lot of changes in how humans (*Homo sapiens*) will continue their existence and evolution. Since numerical simulation has been a highly important technology in the past century and used in almost all engineering-related problems (aerodynamics, aerospace, hypersonic, fossil fuel energy, weather simulation, natural science and environmental engineering, industrial system design and analysis, biological engineering, fluid flows and heat transfer, engine and combustion analysis, and visual effects for film and games), Artificial Intelligence and Machine Learning will provide serious enhancements of numerical simulation.

This book covers Smart Proxy Modeling that has provided the first application of Artificial Intelligence and Machine Learning in the enhancement of numerical simulation. This engineering application of Artificial Intelligence and Machine Learning was successfully originated in 2004 in petroleum engineering (numerical reservoir simulation or NRS) by Intelligent Solutions, Inc., and in 2016 in mechanical engineering (computational fluid dynamics or CFD) by West Virginia University. This book provides details on how Artificial Intelligence and Machine Learning are used for the development of the Smart Proxy Modeling both in NRS and CFD. Numerical simulation is the application of computer science in numerical solution of complex mathematical equations that model multifaceted physical phenomena.

Mathematical modeling of the physical phenomena originated in the 17th century and became more common by the 19th century. Initially, most of the mathematical models were simple enough to be solved using "analytical solution", which means they would provide exact solutions. When composite physical phenomena are to be modeled using mathematical equations, they are no longer simple equations and become quite complex, and therefore, cannot be solved using analytical (exact) solutions. This is how numerical solutions were developed in the 18th and 19th centuries by scientists from Switzerland (Leonhard Euler 1707–1783), France (Joseph-Louis Lagrange 1736–1813), and Germany (Carl Friedrich Gauss 1777–1855).

Numerical solution of highly complex mathematical equation is mainly an approximate solution rather than an exact solution, but most of the time, it works reasonably well. Differences between analytical and numerical solutions of mathematical equations that are used to model physical phenomena were also defined by Creosteanu et al. [2012] in IEEE: "Analytical methods are the most rigorous ones, providing exact solutions, but they become hard to use for complex problems. Numerical methods have become popular with the development of the computing capabilities, and although they give approximate solutions, have sufficient accuracy for engineering purposes".

Today's modern computers were originally initiated in 1936 by Alan Turing at Cambridge University. What is known today as personal computers started in 1975 when Paul Allen and Bill Gates started software development for Altair 8080 that was the world's first minicomputer kit to rival commercial models. As computers became enhanced, numerical solutions started becoming computational numerical solution that would complete the solutions much faster. Starting this process became a tool referred to as computational science.

Computational science that addresses numerical solution of complex multiphysics, non-linear, partial differential equations, is on the forefronts of engineering problem solving and optimization. NRS that is the application of computational science in petroleum engineering, and CFD that is a branch of fluid mechanics which uses numerical analysis and data structures to analyze and solve problems involving fluid flows, are computationally expensive technologies. As the computational numerical simulation started to provide much faster solutions as compared with the solutions prior to the use of computers, it provided scientists and engineers the opportunity to increase the number of cells and grid blocks for numerical simulation and reduce the time-steps.

Developing such details in space (cells and grid blocks) and time (time-steps) would significantly enhance the quality of the solutions generated by the numerical simulation. As the speed of computation of the computers would get enhanced on a regular basis, it would cause enhancement of the details in space (cells and grid blocks) and time (time-steps) by scientists and engineers. The main result of such characteristics has always generated much larger amount of time (footprint) in order to achieve detailed solutions through computational numerical simulation. To reduce the computational footprint of the numerical simulation, few decades ago, scientists and engineers started developing techniques called "proxy modeling".

Proxy models that were developed in the past few decades achieve smaller computational footprint while developing numerical solutions for the complex mathematical equations. However, there are specific issues that are associated with such proxy models. Calling them "Traditional Proxy Models", they simplify the numerical solutions in order to achieve smaller amount of computational footprint. These "Traditional Proxy Models" are known as statistical response surfaces, which are also called response surface modeling (RSM) and reduced order models (ROM). These Traditional Proxy Models attempt to make it practical to use NRS and CFD by addressing their computational footprint.

Smart Proxy Models take advantage of the Artificial Intelligence and Machine Learning solutions to develop replicas of the numerical models. These models are highly accurate, but have very fast response time. The novelty of Smart Proxy Modeling stems from the fact that it is a complete departure from traditional approaches to modeling in the fluid flows and constitutes a major advancement in utilization and incorporation of engineering application of Artificial Intelligence and Machine Learning. Instead of starting with the first principal physics, Smart Proxies are built based on the observation of system behavior, much like how human brain learns.

Smart Proxy of a numerical model is an accurate replication of the numerical simulation model with execution time that is measured in fractions of a second. Using Artificial Intelligence and Machine Learning, Smart Proxy is trained to learn the behavior of the numerical simulator using very large amount of data generated by a single run of the numerical simulation model. Smart Proxy is robustly validated using blind simulation runs.

Shahab D. Mohaghegh, Ph.D.
West Virginia University & Intelligent Solutions, Inc.

About the author

Shahab D. Mohaghegh, a pioneer in the application of Artificial Intelligence (AI) and Machine Learning in the Exploration and Production industry, is Professor of Petroleum and Natural Gas Engineering at West Virginia University (WVU) and the president and CEO of Intelligent Solutions, Inc. (ISI). He is the director of WVU-LEADS (Laboratory for Engineering Application of Data Science).

Including more than 30 years of research and development in the petroleum engineering application of AI and Machine Learning, he has authored three books (*Shale Analytics – Data Driven Reservoir Modeling – Application of Data-Driven Analytics for the Geological Storage of* CO_2), more than 230 technical papers, and carried out more than 60 projects for independents, NOCs, and IOCs. He is an SPE Distinguished Lecturer (2007 and 2020) and has been featured four times as the Distinguished Author in SPE's *Journal of Petroleum Technology* (JPT 2000 and 2005). He is the founder of SPE's Technical Section dedicated to AI and Machine Learning (Petroleum Data-Driven Analytics, 2011). He has been honored by the US Secretary of Energy for his AI-based technical contribution in the aftermath of the Deepwater Horizon (Macondo) incident in the Gulf of Mexico (2011) and was a member of US Secretary of Energy's Technical Advisory Committee on Unconventional Resources in two administrations (2008–2014). He represented the United States in the International Standard Organization (ISO) on Carbon Capture and Storage Technical Committee (2014–2016).

Chapter 1

Artificial Intelligence and Machine Learning

"Artificial Intelligence" is the simulation of human intelligence by mimicking the "Human Brain" for analysis, modeling, and decision-making. "Machine Learning", which includes algorithms for creating Artificial Intelligence, is the science and technology of getting computers to act without being explicitly programed. "Machine Learning" uses Open Computer Algorithms to learn from data instead of explicit programing.

What today is called Artificial Intelligence and Machine Learning started as concepts and ideas in the early 1950s. During the first couple of decades, what is referred to as Artificial Intelligence included rule-based modeling as well as data-driven solutions. The rule-based modeling that ended up being denoted as "Expert Systems" started becoming the only part of what continuously referred to as Artificial Intelligence. This had to do with a book named *Perceptrons* that was written by Marvin Minsky and Seymour Papert in 1969. This book proved that this initial version of the *Perceptrons* (data-driven pattern recognition) does not have the capability of solving non-linear problems. This caused the data-driven pattern recognition to be moved out of the technology called Artificial Intelligence. Some research continued to be performed on data-driven pattern recognition to address non-linear pattern recognition, and finally this technology addressed and solved the negative issues that were identified by Minsky and Papert in their book. The current version of data-driven pattern recognition started to be what it is today, in the mid-1980s.

In the early to late 2000, Artificial Intelligence and Machine Learning were used extensively in some games (Chess, Go, …) and the Internet (Google, Image Recognition, …). In the past couple of decades, the overwhelming majority of applications of Artificial Intelligence and Machine Learning which the society has been exposed to have been used to address non-engineering-related problems. The engineering application of Artificial Intelligence and Machine Learning is quite different from how this technology is used to solve and address non-engineering-related problems, which is also known as human-level intelligence or Artificial General Intelligence (AGI). Engineering domain expertise is an absolute requirement in the engineering application of Artificial Intelligence and Machine Learning.

DOI: 10.1201/9781003242581-1

1.1 MACHINE LEARNING

Currently, Artificial Intelligence is activated through Machine Learning. Machine Learning is the science of making computers to act without being explicitly programed. Machine Learning is done through Open Computer Algorithms and learning from data. "It a set of algorithms that parse data, learn from data, discover patterns in the data, to make predictions or decisions based on what they have learned" [Khan Academy, https://www.khanacademy.org/]. "There are situations in which the learning process takes place by establishing a relationship between input and output labeled data. This learning technique is known as supervised learning. On the other hand, unsupervised learning is a more intuitive approach in which the algorithm can find structure or patterns in unlabeled data through clustering methods, among others" [Khan Academy].

1.2 ARTIFICIAL NEURAL NETWORKS

One of the most common Machine Learning algorithms is the Artificial Neural Network, which is biologically-inspired to mimic the human brain and its learning process. Like information being transmitted from one biological neuron to another via synapse, artificial neurons carry information from one to another through weights and layers. The weights represent the strength of the information being carried while the layers represent the direction in which the information is flowing. As demonstrated in Figure 1.1, an Artificial Neural Network typically includes an input layer, one or more hidden layers, and an output layer.

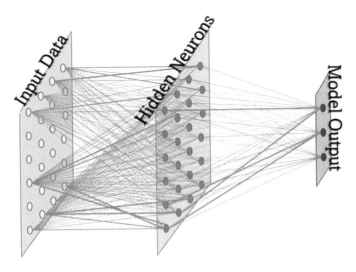

Figure 1.1 Example of an Artificial Neural Network with one hidden layer.

At the input layer, data that is used to train the Artificial Neural Network is initiated and feed-forwarded to the first hidden layer (and then the next hidden layer, if there is more than one hidden layer) and then to the output layer. At the output layer, a predicted value from the Artificial Neural Network is compared to the true target-output value that is used for the supervised training process. The backpropagation algorithm that most of the time is used for training process is implemented to optimize the initial set of weights that could have been generated randomly. Backpropagation algorithm, using the difference between the actual target output and the calculated output through the feed-forward Artificial Neural Network, is used to modify the weights of the connections between all existing neurons.

The feed-forward process is then repeated with the optimized set of weights. Every time the weights are updated, the neural network goes through an iteration or epoch. The predicted value is expected to be closer to the true value after every epoch, suggesting that learning is taking place. From several epochs, the most optimum set of weights is obtained and used for making predictions.

1.3 DEEP LEARNING

The term deep learning mainly refers to a more complex Artificial Neural Network topology, where more than one hidden layer exists. Usually, deep learning models are more successful in the presence of Big Data as they can learn highly complex features, process the data in a faster fashion than previous methods, and are easy to implement.

1.4 FUZZY CLUSTERING

Fuzzy clustering is an unsupervised learning method that groups multidimensional data into smaller subgroups based on a given number of cluster centers. Each data point enclosed in a subgroup has similar characteristics and is distant from every cluster center by a degree. Thus, each data point is assigned a membership to each cluster ranging between 0 and 1. On the other hand, a hard-clustering approach uses a two-valued logic and assigns each data point to a single cluster only.

1.5 FEATURING GENERATION

One of the major characteristics of Domain Expertise during the engineering application of Artificial Intelligence and Machine Learning is Feature Generation. Feature Generation is what engineering domain experts must do with the existing data that is going to be used for the training of Artificial

Neural Network in order to help the training process to be better and faster. One of the characteristics of the Feature Generation includes encoding non-numerical data into numerical data that is meaningful to the learning algorithm. However, other engineering-related encoding of some of the existing data also can be named Feature Generation. In this book, in Chapter 3, examples of Feature Generation are presented.

For example, categorical data that does not have a particular sequence or order can be transformed into integer data by simply assigning an integer value to each of the present categories. This method is known as label encoding. On the other hand, an ordinal encoding approach might be more useful for categories that do have an order. Therefore, an ordinal relationship is created among these categories. However, in situations where such ordering does not exist, it may be misleading to the learning algorithm. Thus, a one-hot encoding approach would be a better fit. In this case, a binary variable is created for each category where values are expressed as either "0" or "1". It is also useful to aggregate the data by taking the average or the sum of a given set of features through grouping operations [Fullmer and Hrenva, 2016].

1.6 PARTITIONING

During the Machine Learning, it is a good practice to partition the data into three different datasets: training, calibration, and validation datasets. While the number of records varies across these datasets, the number of features (columns) remains the same. The training dataset contains most of the data, between 70% and 80% of the entire dataset, from which the learning algorithm will be directly learning.

The calibration dataset is a much smaller portion of the entire dataset and is set aside to provide an unbiased evaluation (during the training process) of the training dataset. The validation dataset is completely independent of the training and calibration datasets, and is used to evaluate the performance of the final trained model. If good results are obtained from the validation dataset, it is likely that the model will perform well on blind (new) datasets [Saripalli et al., 2005].

1.7 NOTE

As it has been noticed, this chapter about Artificial Intelligence and Machine Learning has been quite short and minimal. The main reason has been the fact that the details of engineering application of Artificial Intelligence and Machine Learning, and its differences with non-engineering application of Artificial Intelligence and Machine Learning, are not very simple and short. Furthermore, in this context, the differences between Artificial Intelligence and Machine Learning and traditional statistics should be addressed. Since

Petroleum Data Analytics has been a unique engineering application of Artificial Intelligence and Machine Learning since 1991, major important issues such as Explainable Artificial Intelligence (XAI) and Ethics of Artificial Intelligence (AI-Ethics) have been applied in this technology and must be explained in more detail.

Therefore, instead of a chapter in this book, soon a new book will be published that will go through the details of engineering application of Artificial Intelligence and Machine Learning and Petroleum Data Analytics.

Chapter 2

Numerical simulation and modeling

Modeling physical phenomena is usually done through generation of mathematical equations that would represent the physical phenomena. The degree of complexity of the physical phenomena that is being modeled will determine the complexity of mathematical equations that will be used to represent them. When the mathematical equations that represent physical phenomena are not too complex, they can be solved using analytical (exact) solutions. As the complexity of mathematical equations representing physical phenomena increases, analytical solutions can no longer be used to solve such mathematical equations, unless so many assumptions are made in order to minimize complexity and simplify mathematical equations.

When the complexity of mathematical equations is not simplified enough to make it possible to be solved easily through analytical (exact) solutions, then instead of analytical solutions, numerical solutions must be used to solve them. Numerical solutions require approximation. However, it should be noted that the type of approximation applied to the complex mathematical equation can provide two different approaches to the solutions. If the approximation is made to the essence of the complex physical phenomena, it can simplify mathematical equation so that it can be solved using analytical (exact) solutions. On the other hand, if the approximation is made to the solution, rather than the problem itself, then the solution is known as numerical solution.

The amount and degree of approximation determine the characteristics of the numerical solutions. Such approximations divide the physical phenomena into small pieces in space (cells) as well as small pieces in time (time-steps), so that the linear solutions can be used for each cell in space and time, before combining them to each other to represent the complete model. When the number of cells and time-steps used for numerical solution of complex mathematical equations becomes large, performing numerical solutions using paper and pen becomes quite hard and takes a very long time. When computers became available, numerical solutions became much better to be used to solve complex physical problems. When a numerical solution uses hundreds, thousands, and millions of cells and time-steps, then computers are the only way to generate numerical solutions.

"Numerical Simulation" is the use of computers for obtaining numerical solutions of complex mathematical equations. Numerical Simulation is

DOI: 10.1201/9781003242581-2

a calculation run on a computer following a program, which implements a mathematical model for a physical system. The mathematical equations used to model complex physical phenomena are nonlinear and sometimes include differential equations or partial differential equations. The Numerical Simulations that are addressed in this book (Numerical Reservoir Simulation [NRS] and Computational Fluid Dynamics [CFD]) are some of the most complex, nonlinear, and second-order partial differential equations.

Numerical Simulation started right after the World War II, and the increase and importance of its use became a function of enhancement of computers. As the computers became stronger and faster, use of Numerical Simulation became more enhanced. Among the most common Numerical Simulation methods used for engineering-related problems are Finite Difference Method (FDM) and Finite Element Method (FEM). Numerical Simulation is an important tool required for solving many engineering-related problems such as uncertainty quantification and optimization. The two very common numerical simulation technologies that are currently used in several engineering disciplines are NRS and CFD; their Smart Proxy Modeling will be covered in further chapters of the book.

In many cases, the problems solved with NRS or CFD address very serious issues that need to be solved, optimized, and their uncertainties to be quantified. In order to fulfill such requirements, first and foremost the number of cell (grid blocks) that are used to build numerical simulation model will end up being in tens of millions in order to minimize any simplifications of the solutions and provide reasonable and accurate results for the problem being solved. Since a large number of cells are used in numerical simulations as a requirement to address accuracy of numerical solutions, they create some serious issues with the numerical simulation that has to do with computational time to achieve reasonable solutions.

The most important problem associated with the large number of cells (grid blocks) and time-steps for numerical simulation has to do with the amount of time the solution and deployment of the numerical simulation will require. This is true even when a simple desktop or laptop is not used because of its short computational footprint for such an incredibly large and complex problem. If the computer that is being used for numerical simulation is a desktop or laptop, a single run or deployment of such NRS or CFD will take days and weeks. Once such solutions are done using high-performance computing (HPC), the usual amount of time required would become several hours (about 10 to 20 hours depending on the size and strength of HPC). Different types of HPC systems include different numbers of central processing units (CPUs).

2.1 NUMERICAL RESERVOIR SIMULATION (NRS)

Today, NRS and modeling is the industry standard for comprehensive field studies. Decades of research and development by a large number of engineers and scientists has brought this technology to where it is today. Starting in the late 1960s and early 1970s, advances in the broader fields of technology, such as invention of fast computational machines and development (and adaptation) of algorithms to take maximum advantage of these new computational power resulted in a paradigm shift in reservoir studies from analytical solutions and analog models to more mathematically robust computational models.

This new computational paradigm was able to overcome the mathematical limitations imposed by analytical solutions and was much more realistic than the simple analog models such as Capacitance-Resistance Models (CRMs – introduced to the oil industry in 1943 by Bruce) [Bruce, 1943]. Complex non-linear partial differential equations that govern fluid flow in the porous media were solved numerically at speeds that were unthinkable just a few years before [Aziz and Settari, 1979]. Originally at the time when NRS got started, there were traditional reservoir engineers who did not believe such new technology is a good idea since they were not familiar with the computational technology that was new at the time. Just like any other new technology, numerical simulation had to go through its growing pain struggling against those who failed to understand the change or felt threatened by it. Unfortunately, the same is true today as well, but in a different fashion. Currently there are reservoir engineers who are quite familiar with NRS but do not believe that it would be possible for a new technology to be developed that might be better than NRS. These people are the current traditional reservoir engineers.

Despite the difficulties, which is the character of any paradigm shift, NRS and modeling prevailed. Today, the potential of this technology to solve complex full field problems is hardly contested. It is now a widely accepted technology among engineers and scientists in the oil and gas industry.

In the context of all that has been mentioned, it must be noted that currently there are well-documented shortcomings with NRS and modeling that are controlled and addressed when the study is in the hands of experienced reservoir engineers and modelers. Nevertheless, skepticism on the results provided by NRS and modeling still persists, especially by management teams. Such expressed concerns about numerical simulation models are usually addressed when professionals from different disciplines are involved and cooperate in the full field study.

NRS that is used by operating and service companies requires a solid amount of cost and time. Nevertheless, operating and service Companies that spend such costs for NRS expect returns on their investments. Companies dedicate large amounts of resources to reservoir simulation and modeling. On the one hand, resources are used to secure licenses to commercial simulators or build and maintain in-house simulators and to train engineers and geoscientists in the art and science of numerical reservoir modeling, while on the other hand, even larger amounts of resources are devoted to data collection, performing analyses (core analyses, well tests, etc.) that would be used in the building of geological models that are the foundation of flow models. Yet another important (and expensive) phase of NRS is history matching.

A process that may take months or years (depending on the size of the field) to be completed. Usually, the final result of a successful modeling campaign that takes several years (for a complex full field model) is a history matched, full field reservoir simulation model.

So, what is the Return on Investment (ROI) in the conclusion of a reservoir modeling campaign? ROI in the cases of reservoir simulation and modeling may be calculated in more ways than one. We propose two potential approaches, knowing well that many more can be considered. In one scenario, the amount of money spent on all the items mentioned above that directly contributes to the development and history matching of a reservoir simulation model can be calculated and compared with the number of simulation runs that are made during the life of the model.

In a simplified version of this algorithm performed on a simulation model in a National Oil Company (NOC), we concluded that a single simulation run (the model being examined here includes about 1 million grid blocks and 167 horizontal wells and takes 10 hours to run on a cluster of 12 CPUs) costs about $3,000.

Using a different approach, one may calculate the financial impacts of the decisions made based on the results of such reservoir simulation model and come up with a completely different picture of the ROI of model. It should be noted that unfortunately, such exercises are hardly ever practiced in our industry. Not only calculating reservoir simulation and modeling's ROI is not a common practice, but we also hardly see published studies that bother to look back and see how good the decisions were and measure their economic impact on the companies bottom-line. Sometimes there are good reasons why such practices are not exercised, but we will not get into that discussion here.

2.2 COMPUTATIONAL FLUID DYNAMICS (CFD)

Almost everything that was mentioned in the last section about NRS (complexity, approximation, investment, time, etc.) is also true, to a large extent, and applicable to CFD.

CFD is the numerical simulation of fluid mechanics for solving problems that involve fluid flows. Since fluid dynamics is used to study the evolution of stars, ocean currents, weather patterns, plate tectonics, and blood circulation, CFD can be used for rocket engines, wind turbines, oil pipelines, and air conditioning systems. CFD is also used in research and development of many engineering problems such as aerodynamics and aerospace analysis, hypersonic, weather simulation, natural science and environmental engineering, industrial system design and analysis, biological engineering, fluid flows and heat transfer, engine and combustion analysis, and visual effects for film and games.

It is also applied in technologies such as cavitation prevention, aerospace engineering, heating, ventilation, and air conditioning (HVAC) engineering, electronics manufacturing, and many others. Mechanical and chemical engineers mostly use CFD since it produces quantitative predictions of fluid-flow phenomena based on the conservation laws (conservation of mass, momentum, and energy) governing fluid motion through numerical solutions that is used by computers as numerical simulation.

One of the major issues associated with this important numerical simulation process (CFD) that has the strong capability of addressing a large number of scientific technologies as mentioned earlier, is the amount of time required to make a single run of CFD even on a strong HPC. For example, in the CFD example presented in this book (Chapter 5, Tri-State boiler case study), one single run (deployment) of the CFD on a highly strong HPC would take a full day (24 hours) to provide its answers. This highly large computational footprint is a negative aspect of CFD. When it takes so long for a single run of CFD even using strong HPC, then for making some analysis such as uncertainty quantification or process optimization that requires a large number of CFD simulation runs for its results, the realistic use of CFD for scientific technologies becomes quite hard to achieve. This is the main reason that Smart Proxy Modeling can play an important role in CFD.

Chapter 3

Proxy modeling

Once Numerical Simulation and Modeling became a highly useful technology for analyses, modeling, uncertainty quantification, and optimization of complex physical phenomena, it became a common technology that started to be used by a large number of research and development scientists and engineers throughout the world. When Numerical Simulation and Modeling is applied to complex physical phenomena that require comprehensive and interacted solutions in both space and time, the characteristics and corrective solutions become a function of number of cells and time-steps that are used for their development.

Research and development scientists and engineers soon learned that increasing the number of cells and time-steps for the Numerical Simulation and Modeling is incredibly important for positive and correct solutions. Therefore, in case of important problems related to highly complex physical phenomena whose solutions were being addressed through Numerical Simulation and Modeling, the numbers of cells and time-steps started increasing, and the time for completion of Numerical Simulation became quite large. As the quality of computers increased and calculations became much faster, the required time for the solutions of the Numerical Simulations got reduced. Nevertheless, as the computer calculations became faster, it made research and development scientists and engineers to make larger number of cells and time-steps for the same problem in order to enhance the solutions of Numerical Simulations.

This process caused a new problem since the time required for completion of solutions developed through Numerical Simulations for decision-making became too long to be achieved, and thus using Numerical Simulations to solve highly complex physical phenomena started to become a problem again. Now the problem was not the solution, but how long would it take to achieve a reasonable solution. The long computational time for performing Numerical Simulations created under-utilization of this technology.

This was the reason behind the development of a new technology called "Proxy Modeling". Proxy modeling will be applied to highly complex Numerical Simulations that take a large amount of time to develop a solution, and its objective is to minimize the amount of time required to develop

DOI: 10.1201/9781003242581-3

solutions of Numerical Simulations. Currently, there are two types of "Proxy Modeling":

    a. Traditional proxy modeling
    b. Reduced order model (ROM)
    c. Response surface model (RSM)
    d. Smart Proxy Modeling

The development of traditional proxy modeling started in mid-1950s through mathematical equations and traditional statistics, and that of Smart Proxy Modeling started in 2004 using Artificial Intelligence (AI) and Machine Learning. Smart Proxy Models are developed to address the under-utilization of Numerical Simulation models, which has its roots in the long computational time. Numerical Simulation models are most useful and accurate when they do minimum approximation of a physical problem that is being solved. This minimum approximation of the physical problem enables numerical models to include several millions of grid blocks (cells) in their solutions, and this process causes large computational footprint. Smart Proxy Model replicates the capabilities of Numerical Simulation models with high accuracy at very high speeds. The small computational footprint of Smart Proxy Model makes it possible to fully investigate the capabilities of Numerical Simulation models.

    To the untrained eyes, Smart Proxy Model may resemble traditional proxy models that are common in the numerical modeling circles. However, in this book definition and features of Smart Proxy Model are provided and the reasons behind our argument that Smart Proxy Model is fundamentally different from the traditional proxy models are discussed. The differences between this new generation of Smart Proxy Models with the traditional proxy models will be made clear.

3.1 TRADITIONAL PROXY MODELING

Since Numerical Simulation of complex physical phenomena requires long computational time even when high-performance computers (HPCs) are used for its deployment, it makes certain types of analyses that should be done by Numerical Simulation a hard result to achieve. For example, uncertainty quantification is an essential analysis of many of the physical phenomena that are modeled with Numerical Simulation. It is a fact that performing a realistic uncertainty quantification requires a large number of deployments of Numerical Simulation of physical phenomena. However, when it takes a significant amount of time to deploy even a single run of such a Numerical Simulation, then performing a realistic uncertainty quantification becomes very hard to achieve. Since this has been a negative outcome of the use of Numerical Simulation, several decades ago Proxy Models were developed in order to minimize the computational time of Numerical Simulation.

Given the fact that the definition of numerical solution is the approximation of solution rather than the approximation of problem (not approximating the problem means not minimizing and/or simplifying the complexity of the problem being solved), these traditional proxy models actually move in the opposite direction of the Numerical Simulation by simplifying the problem that is being solved. The objective of the traditional proxy modeling by simplifying the problem (and/or solution) is to minimize the computational time of Numerical Simulation.

This is the major difference between traditional proxy modeling and Smart Proxy Modeling. It is due to the fact that since Smart Proxy Model seriously minimizes the computational time of Numerical Simulation, it makes absolutely no differences in how Numerical Simulation is used to generate solutions for complex physical phenomena.

3.1.1 Reduced order model (ROM)

"Reduced Order Models" (ROM) is one of the traditional types of proxy models that are quite popular. Engineers and scientists have invented many smart ways of reducing the order of Numerical Simulation models in order to overcome the long computational overhead. But of course, as long as one is operating within the realm of a given paradigm, no gain in computational time is without paying a price. The price that is paid by ROM is the accuracy of models.

ROM is defined as an alternative solution to reduce the order of model so it would take less time to provide an answer. The reduction of the order of model may take place by reducing the resolution of model in time and space (larger size grids that result in smaller numbers of grids and larger time-steps that result in smaller numbers of time-steps). This means that mathematical equations of the physical phenomenon that is being solved through Numerical Simulation are reduced through some fundamentally more simplified relationships. One of the most important arguments against ROM is the defeat of the initial purposes behind developing a numerical reservoir simulation (NRS) model, i.e., minimum approximation of the problem being solved.

There are two main ways to reduce the order of a model. One way is to reduce the resolution (both in time and space, but mostly in space). In this approach, the model is grossly up-scaled so much that in some cases the solution for each part of the model approaches the analytical solution with all its shortcomings. Workflows have been developed to increase the static resolution (resolution in space) from a very coarse set of grids in steps in order to find the best middle ground [Williams, 2004].

Another way of developing ROM concentrates on the physics of the problem rather than the space and time resolution of the numerical solution. In the second approach, physics of the model is reduced in order to circumvent the computational time. Many of the most recent examples of such ROM approaches have been applied to numerical modeling of production from shale [Wilson and Durlofsky, 2012]. For example, instead of naturally fracture medium (dual porosity) that increases the computational overhead extensively, some have opted to use single porosity models that are adjusted to act like a dual porosity system. Similar approaches have been adopted for dual permeability models. In these approaches, physical proxies are used to substitute the current understanding of the detailed physics. The final results of ROM approaches are that a different problem is solved, and not the one that originally was the intent of the Numerical Simulation model.

3.1.2 Response surface method (RSM)

The statistics-based proxy models are characterized mainly on the basis of the fact that they deal with the responses that are generated from Numerical Simulation model. That is why they have been dubbed "Response Surfaces". Response surface proxy models require a large number of simulation runs (usually in hundreds, if not thousands) in order to be useful. They suffer from the well-known problems associated with traditional statistics, especially when applied to problems with well-defined physics behind them. One of these well-known problems is the issue of "correlation vs. causality". In other words, simply because two variables correlate, it does not mean that one is the cause of the other. An example that has been documented to demonstrate this point refers to the decline of the murder rate in the United States between the years 2008 and 2014, which demonstrates almost a perfect correlation with the decline of market share of Microsoft's Internet Explorer within the same time period. Although there is a perfect correlation between the two, it is highly doubtful that one has caused the other.

The other well-known problem with all statistics-based approaches is that they impose a pre-defined functional form (mostly polynomial) to the data that is being analyzed or modeled. Of course, one can test a large number of pre-defined functional forms (such as linear, polynomial, exponential, etc.) to finally find the best match, but what if the data representing the nature of a given complex problem does not lend itself to a pre-determined functional form and changes behavior multiple times? Response surfaces are not known to be able to create well-defined and robust input-output relationship between the variables that are crucial in a Numerical Simulation model and the model's responses.

Recently, some new statistics-based proxy models have surfaced that use principal component analysis (PCA) as their core technology. Some recently published works have selected to use different flavors of PCA such as proper orthogonal decomposition (PDO) and polynomial chaos expansion (PCE) in order to develop proxies of numerical reservoir models (Cardoso and Durlofsky, 2010; Chen et al., 2013; He and Durlofsky, 2014; Klie, 2013). These new techniques will ultimately converge to the type of response surfaces that have been around for decades. This is due to the fact that they are being developed within the same computational paradigm, and therefore, it is unlikely for them to provide major breakthrough in this arena. Furthermore, most of these techniques have only been applied to academic problems with small number of cells. The real challenge will surface as they attempt to demonstrate the capabilities of these techniques happen when they are used to build proxies for full field industry-based numerical models with tens of millions of cells (grid blocks).

3.2 SMART PROXY MODELING

"Originally, there was just experimental science, and then there was theoretical science, with Kepler's laws, Newton's laws of motion, Maxwell's equations, and so on. Then, for many problems, the theoretical models grew too complicated to solve analytically, and people had to start simulating. These simulations have carried us through much of the last half of the last millennium. At this point, these simulations are generating a whole lot of data, along with a huge increase in data from the experimental sciences. The world of science has changed, and there is no question about this. The techniques and technologies for such data-intensive science are so different that it is worth distinguishing data-intensive science from computational science as a new, fourth paradigm for scientific exploration" [Bell et al., 2009]. These were excerpts from a presentation by Jim Gray[1] at the National Research Council where he described his vision of fourth paradigm of scientific research.

It must be understood that the application of AI-based modeling of physics in engineering and management is not (and I repeat, is not) a purely statistical approach. The importance of domain expertise in such endeavors cannot be over-emphasized. AI-based modeling of physical phenomena is a fusion of data and knowledge. This gives rise to modeling workflows that are based on data and are guided and constrained by knowledge. It needs to be mentioned here that unlike the use of data in social networking, retail, or even pharmaceutical industry, all AI-based modeling of physical phenomena activities in engineering industry must have the advantage of strong foundation in physics and engineering. In other words, there are sound engineering principles that need to be observed, followed, and respected while modeling almost any process related to the fluid flow, including reservoir engineering and mechanical engineering. When it comes to AI-based modeling of physical phenomena, this turns out to be a curse and a blessing, both at the same time.

It is a curse since there is no precedence for building models based on data in petroleum industry[2]. There are well-understood principles of physics regarding almost all the processes that we encounter in the upstream oil and gas industry, and one needs to follow them in order to build deterministic models. At least that is what the conventional wisdom suggests. Therefore, one better have a good reason (and a thick skin) to charter into the unknown and "non-engineering" territory of data-driven modeling. The amount of resistance put forward by engineers who may feel threatened or are uncomfortable to let their formal education be "toyed" with, can be quite overwhelming.

The blessing is that since there is a set of known physics, it can be used as a validation tool. The AI-based modeling of physical phenomena needs to be thoroughly examined to see if AI-based physical models understand and consequently honor the known physics. This could actually be a strong proving point (validation) for the skeptics[3], since one can demonstrate that it is possible to arrive at the same conclusions and demonstrate that the data-driven model honors all the known physics of a given process.

Emergence of data science as a discipline and "data scientists" was elaborated by the National Science Foundation (NSF) in a report in 2005. Engineering industries have not been and still are not data-centric industries. This is clearly emphasized in modeling engineering problems that our main focus is on physics and mathematics, and data is only referred to (and is accessed) when it is required to serve the needs of the physical model that has been constructed. On the other hand, since engineering managers make crucial decisions on how to move forward through development and optimization, such decisions benefit from facts rather than perceptions. Such facts are data that have been collected through actual measurements. In engineering modeling, we develop governing equations and then solve them numerically using a computer. All this is embedded in the numerical simulators that we use to model. In this approach, the foundation and the ground truth that is treated as a set of facts and is not subject to modification and tuning is our "current understanding" of the physics of a specific physical problem that is being modeled.

Smart Proxy Model is an ensemble of multiple machine learning technologies that are trained to learn and then accurately mimic the intricacies and nuances of the physics of fluid flow in a given hydrocarbon reservoir using (input and output) data generated from the Numerical Simulation model. Smart Proxy Model takes a completely different approach to building proxy models. In this approach, the model is not reduced (neither the physics nor the space-time resolution) like ROM and pre-defined functional forms are not used like response surfaces. Smart Proxy Model is trained (using machine learning algorithms) to learn and accurately mimic the behavior of a comprehensive NRS model. Using Smart Proxy Model, physics and resolution in time and space of the original simulation model are preserved since the data extracted from the Numerical Simulation model is used to train Smart Proxy Model. Since Smart Proxy Model conforms to the system theory, it has an Input-System-Output topology and therefore is not a statistical best-fit of the simulator responses. Using the characteristics of the geological and the flow model as well as the operational constraints that are used in numerical simulator as input and coupling them with the corresponding simulator output, a comprehensive spatio-temporal dataset is generated that includes details of fluid flow that the Smart Proxy Model needs to learn, for a given field.

The assimilated spatio-temporal dataset is used to train and calibrate the Smart Proxy Model. This model is then validated using blind simulation runs and then used for reservoir management and planning purposes. The final result that is the conclusion of the Smart Proxy Model can always be taken back to the numerical simulator for a final run to make sure that the Smart Proxy Model is realistically and accurately reproducing the Numerical Simulation result. The abovementioned steps were implemented during the development and validation of the Smart Proxy Model that is presented in this study. Following is a proposed definition of Smart Proxy Model.

The deployment of the developed (trained, calibrated, and completed) Smart Proxy Model has a small computational footprint, such that thousands of Smart Proxy Model runs can be made in a few minutes using a laptop or desktop workstation. This allows examination of a massive number of scenarios in a short period of time, therefore making it practical to exhaustively search large solution spaces for optimal or near optimal solutions, or to perform Monte Carlo simulations to quantify uncertainties associated with geological models that form the foundation of the numerical flow models.

Smart Proxy Modeling has been studied, developed, and tested to deliver fast and accurate results. Some studies include steady-state and transient problems that have been validated in efforts to replace traditional approaches such as Computational Fluid Dynamics (CFD) and NRS that leave a higher computational footprint and are based on equations. Smart Proxy Models are capable of replicating solution output data from CFD and NRS or any other Numerical Simulations.

NOTES

1 Legendary American computer scientist who received the Turing Award in 1998 "for seminal contributions to database and transaction processing research and technical leadership in system implementation".
2 Please note that Decline Curve Analysis (DCA) is not being considered as a data-driven modeling exercise here. DCA is merely a curve fitting exercise.
3 Of course, it is needless to say that there are some (and I have met a few in my time) who are not skeptics but rather too religiously attached to a certain way of thinking, attempting to convince these individuals of a world beyond their current beliefs, sounds to them, as heresy. They cannot fathom that there are other ways of solving problems. Efforts to convince such individuals is simply a waste of time.

Chapter 4

Smart Proxy Modeling for numerical reservoir simulation

Let's start with a two-part claim and then try to prove each part of the claim: (a) Numerical reservoir simulation (NRS) models are currently under-utilized and (b) the technology covered in this chapter of the book provides the means to substantially increase the utilization of NRS. Of course, it is understandable and conceded that not all NRS models are created equal, and there are models built by individuals that have little utility to begin with. However, the topic covered in this book targets models (NRSs) that are built and history matched by engineering professionals who adhere to the highest standards in the industry. Such models use the state-of-the-art tools and techniques in reservoir simulation and modeling.

In other words, NRS models can offer much more than they are currently being used for. But how do we substantiate this claim? We must mention that one of the motivations behind writing this chapter of the book is to substantiate this two-part claim, and hopefully, by the end of this book, when all the facts are presented, there will remain little doubt about its substantiation. It is a fact that when an NRS model takes tens of hours for a single run (even on most advanced parallel computing architectures), not much can be expected from it. Any serious full field study requires a large number of simulation runs to either search a very large solution space or to quantify uncertainties that are inherent in the geological model.

To address this issue, scientists and engineers in petroleum and other industries have invented smart techniques to reduce the computational footprint of numerical simulation models. This usually happens in one of the two forms. Either speed up the calculations or calculate less. To speed up the calculations, some operators have adopted fast computing techniques, parallelization of the algorithms, or simply brute force, bundling a large number of CPUs. In other word, they have chosen to decrease the time required for a simulation run by turning the wheel faster and faster. This sometimes works, but up to a point.

Furthermore, as you introduce more computing power to your armament, the possibility of faster models makes modelers build models with higher resolution and therefore models start getting so large that the final outcome remains the same, a race that has been tried before in other industries with

DOI: 10.1201/9781003242581-4

predictable outcomes. The two potential solutions mentioned in Chapter 3 of the book are reduced order models (ROM) and response surface methods (RSM).

In the potential solutions that were numerated above ROM and RSM, one thing remains constant. These solutions use the same general paradigm (computational mathematics) to address the problem that was caused by using this paradigm in first place. No wonder it looks like we are going on and on in a circle. In order to snap out of this circle, we need to shift our paradigm. In other words, if the paradigm of building an NRS model is to understand the physics, develop governing mathematical equations for the underlying physics and then use numerical and computational techniques to solve the mathematical equations. At that time, maybe these same steps (approach) should not be used in solving the problem that they have caused, i.e., large computational footprint.

Smart Proxy Modeling of NRS can be developed in two ways: Well-based Smart Proxy Modeling and cell-based Smart Proxy Modeling. The characteristics of the type of Smart Proxy Modeling that is developed for NRS have to do with the objectives that must be accomplished.

4.1 WELL-BASED SMART PROXY MODELING

Well-based Smart Proxy Models can be developed based on the information generated by the NRS for all the production and injection wells that have been modeled in the simulator. Since fluid flow in porous media is physically modeled by diffusivity equation and it is the main characteristic of the NRS, activities and results of the production and injection wells are a function of space and time. Therefore, all the space and time data used in the NRS for each well must be used as input during the development of the well-based Smart Proxy Model.

A well-based Smart Proxy Model is able to generate the production from each well as a function of time for any possible space and time characteristics that can be used in the NRS. The main key behind the well-based Smart Proxy Model is that tens of millions of simulation runs can be made in a short period of time in order to be able to quantify the uncertainties associated with all space and time data that will be used in the modeling process as well as in production optimization of wells that are planned to be drilled in different locations in the field that has been modeled using NRS.

The next two actual case studies that have been performed in United Arab Emirates (ADNOC) and Saudi Arabia (ARAMCO) demonstrate how well-based Smart Proxy Model have been developed and what kind of results were generated. The first well-based Smart Proxy Model for ADNOC was performed on a history matched NRS to optimize oil production and the second well-based Smart Proxy Model for ARAMCO was performed on a greenfield (not history matched) NRS to identify the best places to drill oil production wells.

Case study: Oil production optimization in a mature field – ADNOC

This prolific, mature asset that includes more than 160 production wells has been the subject of peripheral water injection for many years to maintain pressure and help displace oil toward the production wells. Production was restricted to 1,500 barrels of fluid per day per well to avoid excessive water production as well as pulling oil from an over-produced overlaying prolific reservoir. In 2005 a reservoir management study was commissioned to evaluate the impact of rate relaxation on this asset. The objective was to explore the likelihood of increasing oil production from the asset while minimizing the possibility of increasing the water cut.

The study was performed using an existing history matched reservoir simulation model. To maximize the utility of the numerical model, a well-based Smart Proxy Model was developed and used for the study. Upon completion of the study (during the second quarter of 2005), the well-based Smart Proxy Model ranked all the wells, based on the probability of success that was defined by the management as producing large volumes of incremental oil with minimal incremental water production (maintaining or reducing water cut), once the rate relaxation program is implemented. In January 2006, the management issued permission for rate relaxation of 20 wells in the field.

In this section of the book, results and consequences of the field management decisions are evaluated using actual field data that are provided for more than five years after their implementation. The well-based Smart Proxy Model that had been used in this project helped reservoir and production managers, engineers, and modelers make the most of the tools that are at their disposal to make more informed field management decisions. In this section of the book, it is demonstrated that using the right tools and strategies, skepticism about reservoir simulation models can be addressed effectively and can result in highly successful practices.

In 2005, ADCO's management (ADCO – Abu Dhabi Company for Onshore Oil Operations is currently ADNOC – Abu Dhabi National Oil Company) decided to embark on a new journey and test drive a completely new and non-traditional reservoir management technology. This was indeed a bold move. This technology was applied to a mature field with 167 production wells (Figure 4.1). There was a notion by some of the shareholders that wells in this asset have the capacity to produce oil at higher rates than what are currently being allowed. The restrictions imposed on all wells included 1,500 barrels of liquid per day per well, not exceeding 250,000 barrels of oil per day for the entire asset. Large amounts of water injection along with highly complex geology (this is a naturally fractured carbonate reservoir) indicated omnipresence of the threat of high

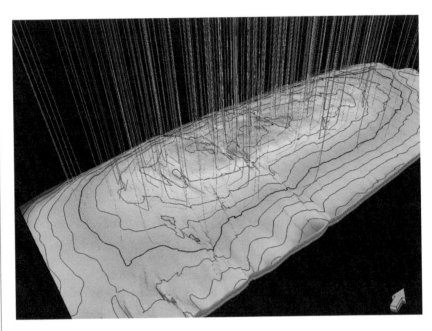

Figure 4.1 Structural map of the reservoir including a large number of wells. Image extracted from the full field numerical simulation model.

water cut for each well. This included losing the wells and bypassing oil banks due to water fingering when wells were pulled at high DF: differential pressure.

Convincing the management to lift some of the rate restrictions (open the choke on some wells and allow higher oil productions) was not an easy task since the danger of high water cut (loosing wells and leaving bypassed oil behind) was quite imminent. Furthermore, there was no indications of which wells should be subjected to a potential rate relaxation program since reservoir simulation runs would provide results based on the imposed production constraints and would not be able to examine all possible scenarios in a practical time span. This would get even more complicated when uncertainties associated with the geological model would be considered.

The history matched NRS model for this asset includes more than one million grid blocks and has a runtime of about ten hours when implemented on a cluster of 12 parallel CPUs. The project team decided that in order to be able to make reliable reservoir management decisions, all possible combinations of changes in operational constraints (different choke sizes for a large number of combination of wells) must be examined while the uncertainties associated with the geological model is quantified. Such exercise required a massive number of simulation runs (more than a million). Knowing that systems developed based

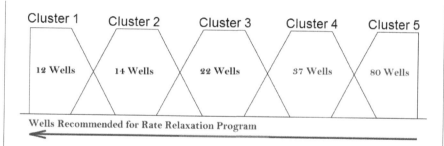

Figure 4.2 Wells in the asset were divided into five clusters.

on Machine Learning have very small computational footprint and can be run almost instantaneously, the idea of using Machine Learning to build a replica of the numerical simulation model was proposed. At first, it sounded like an impossible undertaking, but eventually such a model was generated, and the resultant proxy model was called Smart Proxy Model. Details of how the Smart Proxy Model was developed and tested for accuracy and then used for uncertainty quantification has been covered in multiple papers [Mohaghegh, 2006a, 2006b, 2006c].

As was mentioned in those papers, upon the completion of the Smart Proxy Model runs and analysis of the results, wells in the asset were ranked based on the probability of success (high incremental oil production and low water cut) in a potential rate relaxation program. The wells that ranked highest would have a higher probability of success, and the wells that were ranked low were not recommended as a candidate for rate relaxation. Furthermore, the ranked wells were divided into five clusters as shown in Figure 4.2.

Wells with highest to lowest probability of success were classified in Clusters 1 through 5. Success was defined as high incremental oil production along with minimum increase in water production. Wells in Clusters 1, 2, and 3 were labeled as "Definitely a Candidate", "Candidate", and "May be a Candidate", respectively. Furthermore, it was mentioned that wells in Clusters 4 and 5 have the tendency of producing large amounts of water such that they should not be considered as candidates for rate relaxation (wells in Clusters 4 and 5 were labeled as "Not a Candidate" and "Definitely Not a Candidate"). Figure 4.3 shows the relative location of the wells of different clusters in the field.

Using the Smart Proxy Model, we were able to make thousands of runs in seconds replicating the numerical simulation model's behavior. For example, considering the results used in the analyses, plots of cumulative oil production and water cut as a function of time and change in the cap rate (choke size) were made for all wells. Figure 4.4 shows two examples of such plots (please note that these are not response surfaces, but direct output of the Smart Proxy Model).

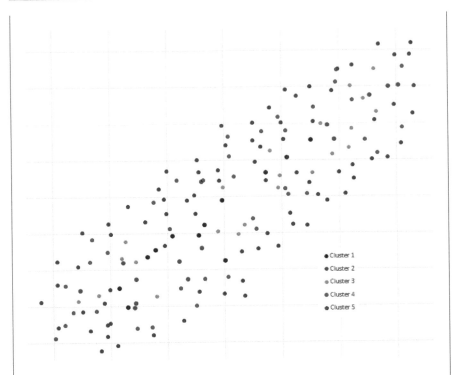

Figure 4.3 Locations of wells in the five clusters based on SRM analyses.

In this figure, cumulative oil production is shown on the left 3D plot and water cut is the 3D plot on the right of each part of the figure.

The top plots in Figure 4.4 show the result for the well that was ranked number 1 as a candidate for rate relaxation (belonging to Cluster 1). It was predicted that this well has the capacity to produce at cap rates up to 4,500 barrels per day without any danger of high water cut. On the other hand, the bottom plots in Figure 4.4 show the well that was ranked number 75 (belonging to Cluster 4). The figure shows that although this well has the capability of producing large volumes of oil, it has the tendency of having water cuts as high as 75%, and therefore should not be considered as a candidate for rate relaxation.

Given the fact that a large number of Smart Proxy Model runs could be made in only a few seconds/minutes, performing Monte Carlo simulation to quantify uncertainties associated with the geological model becomes a practical undertaking. Figure 4.5 displays such an example. In this figure, results of Monte Carlo simulation performed on the well that was ranked number 2 (belonging to Cluster 1) is shown. All reservoir characteristics for this well (formation thickness, porosity, permeability, etc.) were assigned ranges of values in the form of

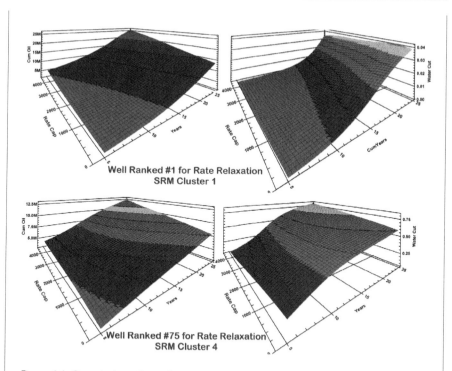

Figure 4.4 Cumulative oil production and water cut as a function of time and cap rates for the well ranked number I (top plots) and number 75 (bottom plots) by Smart Proxy Model analysis.

triangular distribution instead of single values that were in the history matched numerical simulation model.

In the well shown in Figure 4.5, the time was set to five years and the cap rate was fixed at 3,000 barrels per day. The Smart Proxy Model was executed for 1,000 times and the results (cumulative oil production and water cut at the end of fifth year) were plotted in the form of probability distribution function. These calculations and generation of plots shown in Figure 4.5 took about 6 seconds.

This figure shows that cumulative oil production for this well has values of 1 million, 2 million, and 4.5 million barrels for P90, P50, and P10, respectively. Furthermore, P90, P50, and P10 values for the water cut are 1%, 3%, and 20%, respectively. It is clear to see why this well was selected as a candidate for rate relaxation. Please note that results shown in Figure 4.5 represent only one out of a large number of production scenarios that were examined.

Upon completion of this study and presentation of results to the management, it was decided to allow the commencement of a rate relaxation program in this asset. From January 2006 through February 2007, 20 wells in total were approved by the management for the rate relaxation program (seven wells

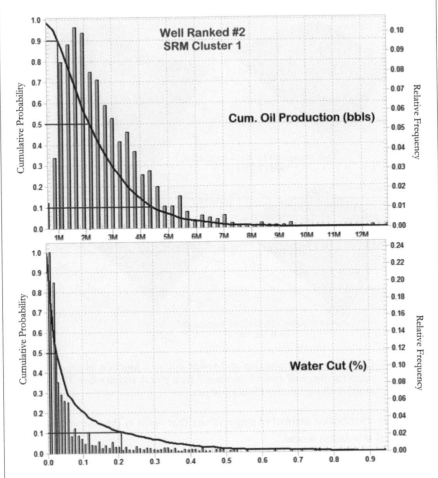

Figure 4.5 Monte Carlo simulation results for cumulative oil production and water cut for the well ranked number 2 by Smart Proxy Model analysis.

in January 2006, three wells in November/December 2006, and ten wells in February 2007) to see the impact of producing more oil on water production.

This program would allow higher amounts of oil production from the asset while keeping the well count constant (increasing total asset output without drilling new wells and accessing new reservoir). List of wells and their association with the classifications made by the Smart Proxy Model analysis (in 2005) are shown in Figure 4.6 while the locations of all wells that were selected to participate in the rate relaxation program are shown in Figure 4.7.

It is interesting to note that (a) the wells being tested for rate relaxation are geographically spread throughout the field. This way the reservoir management could sample the impact of well location (based on the structure of the reservoir)

	January 2006	November 2006	February 2007	Total
Cluster No. 1	2	1	1	4
Cluster No. 2	1	0	1	2
Cluster No. 3	1	1	4	6
Cluster No. 4	1	1	2	4
Cluster No. 5	2	0	2	4

Figure 4.6 List and time of wells in different clusters approved for rate relaxation.

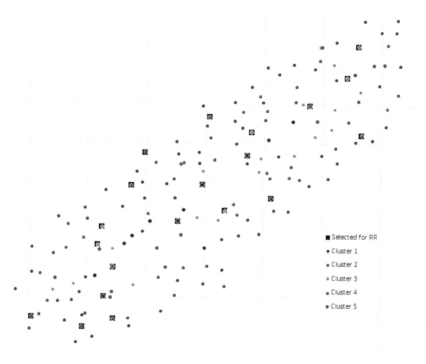

■ Selected for RR
● Cluster 1
● Cluster 2
● Cluster 3
● Cluster 4
● Cluster 5

Figure 4.7 Location of the wells approved for the rate relaxation program between January 2006 and February 2007.

on water production, and (b) by spreading the candidate wells throughout the field, distribution of wells among all the five clusters was achieved. In other words, all predictions made by Smart Proxy Model were being tested.

In mid-2010, after three to four years of production from the wells that were the subject of the rate relaxation program, the management decided to perform a "look-back" study in order to assess the impact of the decisions made in 2006 and 2007. In this study, field measurements of oil and water production (from 2006 through 2010) were compared with the predictions made by the Smart Proxy Model study in 2005. As mentioned above, since wells from all five clusters were the subject of the rate relaxation program, this look-back study would comprehensively substantiate the value of Smart Proxy Model, based on its prediction, versus actual field measurements.

A summary of oil production and water cut for each of the 20 wells and their associated clusters is presented in Figure 4.8. This figure includes three bar charts in a column that share the same x-axis. The common x-axis in all the charts identifies the wells and the cluster they belong to. For example, the first four are identified as 1a, 1b, 1c, and 1d. This simply means that each of the first four bars in this figure (all three charts) reflects the corresponding values for the four wells in Cluster 1, and so forth.

In this figure, the top bar chart shows the maximum water cut for each of these wells prior to the rate relaxation program. The y-axis of this chart is between 0 and 14% water cut. Please note that all the 20 wells that were selected to participate in the rate relaxation program had very low water cut. There are two items of interest in this top chart. First item is that almost all of the 20 wells had low water cut prior to the program. This shows that while selecting the wells for the rate relaxation program, the management made sure to select wells that are least susceptible to cutting water. The second item is that among the selected wells, three wells have the highest water cut prior to rate relaxation. These are well number 1b (belonging to Cluster 1) with more than 13% water cut prior to the rate relaxation program, well number 2b (belonging to Cluster 2) with more than 9% water cut prior to the rate relaxation program, and well number 3a (belonging to Cluster 3) with more than 6% water cut prior to the rate relaxation program.

Interestingly, these three wells were among the wells recommended by Smart Proxy Model as rate relaxation candidates. They were predicted by Smart Proxy Model analyses to be high performers after implementing rate relaxation (which looks and sounds pretty counter-intuitive). In other words, the wells that on the surface looked to be the worst (in terms of water cut) were predicted to be among the best performers by the Smart Proxy Model analysis. On the other hand, all the eight wells belonging to Clusters 4 and 5 (that were predicted by the Smart Proxy Model to perform poorly as far as water cut is concerned and therefore were NOT recommended as candidate for rate relaxation) had water cut values ranging from less than 1% to less than 3%. Therefore, water cut prior to the rate relaxation program seemed to go against the predictions made by the Smart Proxy Model.

The bar chart in the middle (Figure 4.8) shows the maximum water cut achieved by each of the wells after the rate relaxation program (until the date of the look-back study in mid-2010). The y-axis of this chart is between 0 and 55% water cut. This chart shows that well number 1b that started at more than 13% water cut, ends up having a water cut of less than 4% after the rate relaxation.

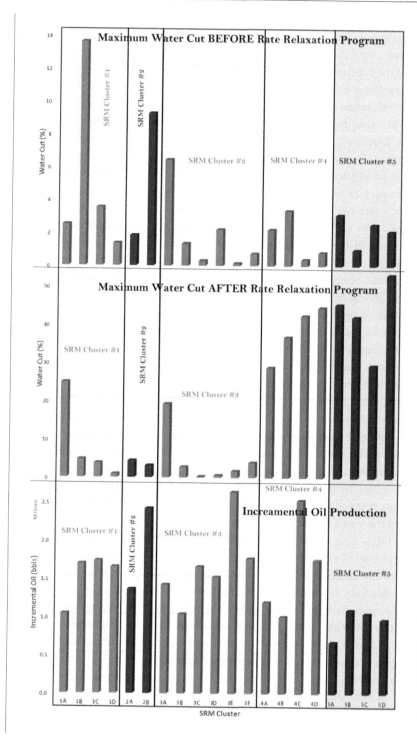

Figure 4.8 Field production history before and after rate relaxation program indicating incremental oil production and water cut.

The decrease in water cut simply indicates that the increase in oil production was much more than the increase in water production since the commencement of the rate relaxation program, therefore reducing the water cut to 11%. Similar results are achieved for well number 2b (more than 9% water cut before the rate relaxation program reduces to less than 3% after the program).

On the other hand, wells in Clusters 4 and 5 behave very differently from wells in Clusters 1, 2, and 3. Wells in Clusters 4 and 5 that started the rate relaxation program with an average water cut of below 3%, show water cuts in excess of 30% and in some cases as high as more than 50% only in the three to four years of production since the rate relaxation program started. The trend of water cut increase in these wells promises even higher water cuts as the production with relaxed rates are continued. These two bar charts clearly demonstrate that as far as the water cut is concerned, predictions made by the Smart Proxy Model have proven to be quite accurate.

It is always obvious that once the rate relaxation program is initiated, more oil can be produced from every single well in this asset. This was actually pointed out by the management on many occasions. The management always indicated that there was no doubt that all wells in this asset are capable of producing more oil (a fact that is clearly demonstrated by the bottom bar chart in Figure 4.8), but the catch always was about the water production. The key was to identify the wells that would not be jeopardized by high water cut.

Figures 4.9–4.14 display detail field measurements for six of the wells in the rate relaxation program. The first three figures (Figures 4.9–4.11) are examples of wells that were part of Smart Proxy Model recommendations to be rate relaxed, and the next three figures (Figures 4.12–4.14) are examples of wells that were identified as not being candidates for the rate relaxation program, and therefore, not recommended by the Smart Proxy Model study. In these figures, the x-axis is time (date), the y-axis on the left is cumulative production in barrels, and the y-axis on the right is water cut in percentages.

The black dots (continuously increasing) show the field measurements of cumulative oil production while the green dots show the field measurements of cumulative water production. The shades of blue are the water cut. Commencement of rate relaxation program is indicated by a red downward arrow. A gray dash-line arrow shows the extrapolation (projection) of the cumulative oil production if the well would not have been subjected to rate relaxation and would have continued in its previous path of production.

Therefore, the difference in cumulative oil production at the end of the actual and the projected cumulative oil production may be considered as the incremental oil production attributed to the rate relaxation program. For example, in Figure 4.9 (that is Well number 1C – This well was ranked number 3 in the

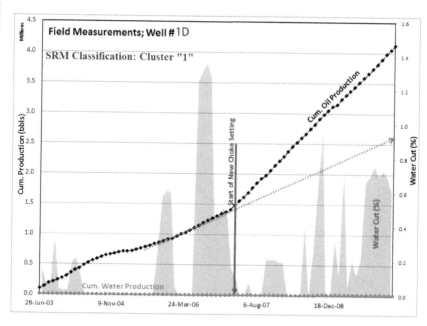

Figure 4.9 Cumulative oil and water production as well as water cut before and after the rate relaxation program. Well number 1D.

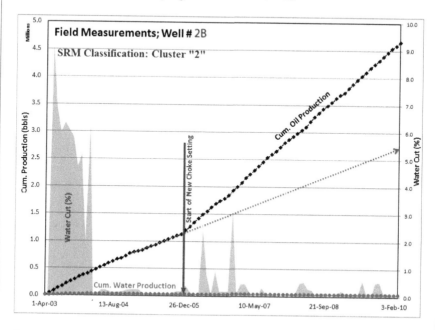

Figure 4.10 Cumulative oil and water production as well as water cut before and after the rate relaxation program. Well number 2B.

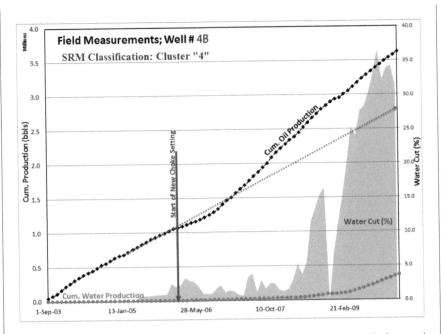

Figure 4.11 Cumulative oil and water production as well as water cut before and after the rate relaxation program. Well number 4B.

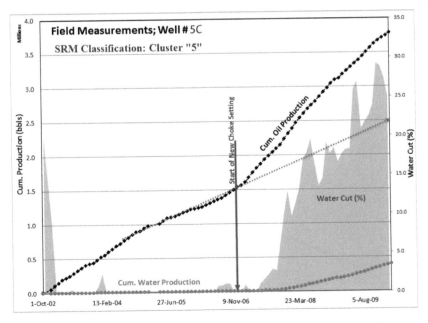

Figure 4.12 Cumulative oil and water production as well as water cut before and after the rate relaxation program. Well number 2C.

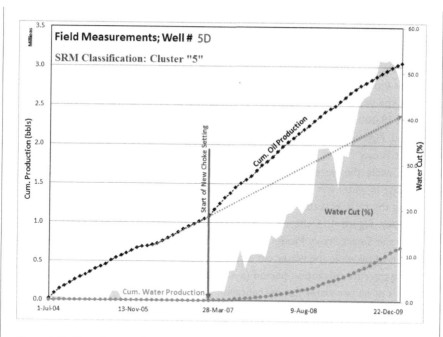

Figure 4.13 Cumulative oil and water production as well as water cut before and after the rate relaxation program. Well number 5D.

ISI Cluster No.	Total Number of Wells	Revenue
Clusters 1 & 2	6	$673,467,997
Cluster 3	6	$700,440,438
Clusters 4 & 5	8	$532,905,117
	TOTAL	1,906,813,552

Figure 4.14 Total revenue from the wells that participated in the rate relaxation program based on an average oil price of $75.00 per barrel.

list of candidates by the Smart Proxy Model), the projected cumulative oil production is about 3.1 million barrels while the post rate relaxation cumulative oil production is about 4.1 million barrels. This means that only from this one well in Cluster 1, one million barrels of incremental oil was produced (in the few months since the start of the program) with no impact on the water cut.

Figure 4.10 shows another Cluster 1 well that was ranked number 11 by the Smart Proxy Model analysis. This well has produced extra 1.5 million barrels of oil with no water cut increase. Figure 4.11 shows a Cluster 2 well that was ranked number 18 by the Smart Proxy Model analysis. This well has produced extra 1.6 million barrels of oil with no water cut increase.

Figure 4.12 shows a Cluster 4 well that was ranked number 71 by the Smart Proxy Model analysis. This well has produced extra 0.6 million barrels of oil while the water cut increased from 2% to 35% (a 33% increase in water cut or about 18,000 barrels of oil per water cut percentage) during the rate relaxation period. Figure 4.13 shows a Cluster 5 well that was ranked number 95 by the Smart Proxy Model analysis. This well has produced extra 1.2 million barrels of oil while the water cut increased from less than 1% to about 30% (a 29% increase in water cut or about 133,000 barrels of oil per water cut percentage) during the rate relaxation period.

Finally, Figure 4.14 shows another Cluster 5 well that was ranked number 92 by the Smart Proxy Model analysis. This well has produced extra 0.7 million barrels of oil while the water cut increased from less than 1% to more than 50% during the rate relaxation period (a 49% increase in water cut or about 14,000 barrels of oil per water cut percentage).

Figure 4.15 shows that at an average price of $75 per barrel, the eight wells in Clusters 4 and 5 (not recommended by the Smart Proxy Model analysis as candidate wells) have resulted in approximately $533 million revenue, while many of them produced so much water that may had to be shut-in (this may be misleading since usually an increase in water cut may be attributed to large volumes of oil in-place that have been by-passed and may never be recovered.) On the other hand, the 12 wells in the first three clusters (that were identified as candidates for rate relaxation by Smart Proxy Model) have generated close to $1.3 billion revenue with negligible water cut increase such that none of these wells are in danger of being shut-in.

Case study: Potential oil production from green fields – ARAMCO

Field "R" is an onshore field that is produced from carbonate reservoirs in anticlinal traps consisting of an upward shoaling sequence of marine carbonate capped by anhydrite. The formation consists primarily of limestone and dolomites with reservoir heterogeneities characterized by the existence of high permeability streaks, faults, and fractures. The review of logs suggests that reservoir characteristics can vary significantly over relative short distances.

Oil-water contact was interpreted for this reservoir. The API gravity of the oil varies from 24° to 32° in the oil-bearing zones, and the crude viscosity varies from 1.5 cP to 4 cP. The reservoir is initially undersaturated and is currently undeveloped.

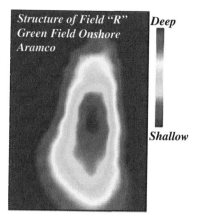

Figure 4.15 Structural map of field "R" in Saudi Arabia.

Reservoir simulation model was constructed for this field primarily for field development planning purposes using the in-house massively parallel processing simulator. The model is three-phase, three-component (oil-water-gas), and single-porosity single-permeability, although no free gas in the reservoir is expected due to planned pressure maintenance by water injection. The simulation model for field "R" consists of 1.4 million active cells with an aerial grid size of 250 m × 250 m.

The model includes 40 vertical production wells and 22 vertical water injection wells; the run time of the model is approximately 12 minutes on a cluster of CPUs. Figures 4.15 and 4.16 show some of the reservoir characteristics of field "R". Due to the limited number of wells that have been drilled in this field, considerable uncertainties exist in reservoir description as well as in the

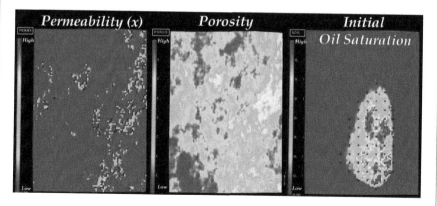

Figure 4.16 Permeability (in x direction), porosity and initial oil saturation of field "R" in Saudi Arabia.

Table 4.1 Nine simulation runs designed for the development of the
Smart Proxy Model for field "R"

Run number	Bottom-hole pressure (psi) for all the wells in the field	Maximum liquid rate (bbls/d)
1	500	10,000
2	500	15,000
3	1,000	10,000
4	1,500	10,000
5	1,500	15,000
6	1,500–1,000–500 (variable in steps)	10,000
7	1,500–1,000–500 (variable in steps)	15,000
8	1,500–500 (variable continuous)	10,000
9	1,500–500 (variable continuous)	15,000

engineering data. Performing uncertainty/risk analysis is critical to realistically assess the potential of field development.

As mentioned earlier, a small number of simulation runs is sufficient for the development of the Smart Proxy Model. A total of nine simulation runs were designed for the development of the Smart Proxy Model for field "R". Table 4.1 summarizes the operational constraints that were used for each of the nine simulation runs. In this table, it is shown that the bottom-hole pressure (BHP) and the maximum liquid rate in each run were varied within the expected operational ranges. These operational constraints were imposed on all the producing wells in the field. In five of the simulation runs, the BHP was kept constant for the entire 20 years of production, while in four of the simulation runs, the BHP was varied as a function of time. "In steps" and "continuous" modifications of BHP are shown in Figure 4.17.

Smart Proxy Models are developed using data extracted from simulation runs. Therefore, the first step in any Smart Proxy Model project starts with developing a representative spatio-temporal database. The extent at which this spatio-temporal database actually represents the fluid flow behavior of the reservoir that is being modeled, determines the potential degree of success in developing an accurate model.

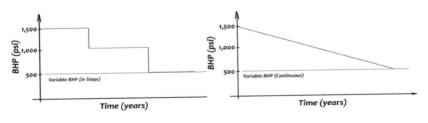

Figure 4.17 Variable BHP constraints in steps and continuous use in the simulation runs.

The term spatio-temporal defines the essence of this database. It is inspired from the physics that controls this phenomenon and is described by the diffusivity equation. The main objective of modeling a reservoir is to be able to know the value of pressure and saturation at any location in the reservoir at any time. Therefore, data collection, compilation, organization, and processing must be performed with such needs in mind.

An extensive data mining and analysis process should be conducted at this step to fully understand the data that are housed in this database. The data compilation, curation, quality control, and preprocessing are the most important and time-consuming steps in developing the Smart Proxy Model. "Curse of Dimensionality" is one of the issues that is associated with Smart Proxy Model and must be handled eloquently during this step of the process. Proper handling of this important issue can make or break the entire modeling process.

The main ideas behind the development of the spatio-temporal database are:

Dynamic allocations of reservoir volume to each well, using the modified Voronoi Graph Theory, along with dynamic and operational based up-scaling of the geological model, accurate coupling of the information generated in step (a) with the time-based production and injection activities, and association of each well with its corresponding offset wells (injectors and producers both in space and time) are used to generate a cohesive record of an event that adequately represents the fluid flow in the reservoir.

One of the most important steps in the development of the Smart Proxy Model is the identification of key performance indicators (KPIs). The spatio-temporal database that was developed in the previous step includes a very large number of parameters that need to be analyzed and possibly included in the predictive model. It is a fact that not all of the parameters have an equal impact on the oil and gas production throughout the reservoir. Using a large number of input parameters in developing a predictive model will result in a system with serious tractability issues. Therefore, it is very important, and even vital to the success of the training, matching, and validating of the Smart Proxy Model to be able to efficiently identify the KPIs of a given model.

Results of the KPI analysis are summarized in Figures 4.18 and 4.19. Keeping in mind that in the spatio-temporal database, the reservoir characteristics of each of the wells along with their associated offset wells have been compiled, the KPI analysis can identify the impact of each of these parameters on the production from each well.

In Figure 4.18, all the reservoir properties that are associated with the well itself (not including the properties of its offset wells) are compared with one another. The length of each bar is indicative (comparatively) of the impact that each property has on the oil production. Also, as seen in this figure, it is clear

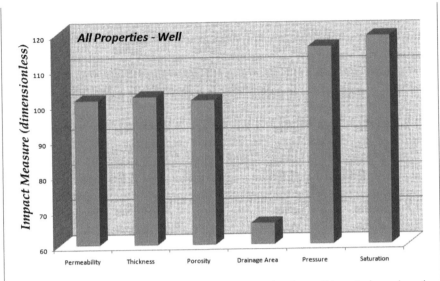

Figure 4.18 Impact of the reservoir characteristics (at the well location) on the oil production.

that the initial oil saturation in the reservoir has the highest impact on the oil production followed by initial pressure. The three reservoir characteristics, porosity, permeability, and formation thickness seem to have equal (but less than saturation and pressure) impact on the oil production followed by drainage area.

In Figure 4.19, each of the reservoir characteristics has been evaluated independently and the corresponding reservoir property at the well location has been compared with the two closest offset wells. The processes of

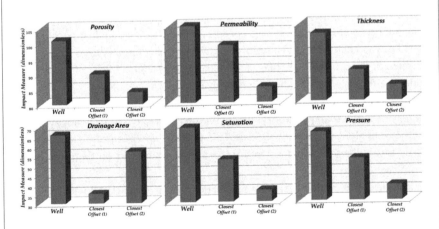

Figure 4.19 Comparing the impact of reservoir characteristics on the well location with those from closest offset wells (numbers 1 and 2).

building (training) the Smart Proxy Model and matching its performance with that of the reservoir simulation model are performed simultaneously. During these processes, the Smart Proxy Model is trained to learn the reservoir model and the fluid flow behavior in the specific reservoir simulator being modeled. The spatio-temporal database developed in the previous step is the main source of information for building and matching the Smart Proxy Model. Please note that the Smart Proxy Model may be a collection of several models that are trained, calibrated, validated, and finally used in concert to generate the desired results.

Issues that must be taken into consideration at this step of modeling include the status of the reservoir (modeling a green field and a brown field are completely different), the purpose of the model (Smart Proxy Models developed for history matching purposes and those developed for predictive analysis purposes), and the objective of the model (modeling pressure and saturation changes in the reservoir vs. modeling injection and production behavior at the well or coupling both in one model). Each of the above-mentioned issues determines the nature of the tools and the strategies that are used in developing a successful Smart Proxy Model.

It is of the utmost importance to have a clear and robust strategy for validating the predictive capability of the developed Smart Proxy Model. The model must be validated using completely blind simulation runs that have not been used, in any shape or form, during the development of the Smart Proxy Model. Nevertheless, it is important to select the blind runs such that they fall within the ranges (both static and dynamic) that were used during the training of the Smart Proxy Model.

Both training and calibration data sets that are used during the initial training and matching of the Smart Proxy Model are considered non-blind. Some may argue that the calibration – a.k.a. testing data set – is also blind; this argument has some merits but if used during the development of the Smart Proxy Model, it may compromise the validity and predictability of the model, and therefore, such practices are not recommended.

During the training and calibration of the Smart Proxy Model for field "R", data from the nine simulation runs that were organized in the spatio-temporal database was used. The training and calibration included 65% and 10% of the data in the spatio-temporal database, respectively. The calibration data set is used as a blind data set to make sure that the Smart Proxy Model is not over-trained. Figure 4.20 is a snapshot of the training of the Smart Proxy Model.

The remaining 25% of randomly selected data (even during the training and matching process) was left out as a blind data set for validation of the Smart Proxy Model. Since this 25% is randomly selected from within the spatio-temporal

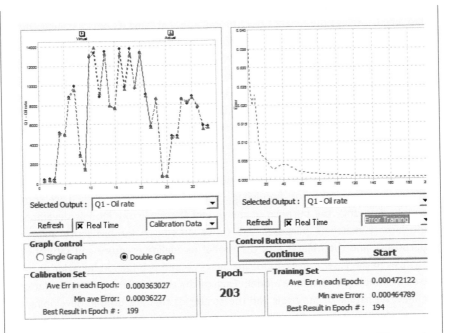

Figure 4.20 Smart Proxy Model of field "R" during the training process. The graph on the left shows the performance of the Smart Proxy Model on the calibration data set while the graph on the right shows the change in the error of the training data set.

database, it does not represent a cohesive simulation run from start to finish. Therefore, we do not consider this 25% of randomly selected data as a measure for the validation of the Smart Proxy Model. Nevertheless, if the Smart Proxy Model does not perform satisfactorily on this 25% of randomly selected data, then the chances of it performing well on a completely blind simulation run would be minimal.

Figure 4.21 shows the performance of the Smart Proxy Model on the 25% of randomly selected data from within the spatio-temporal database. It is clear from this figure that the Smart Proxy Model displays a strong performance in predicting the oil rate for a blind data set.

Figures 4.22–4.24 show the comparison of the Smart Proxy Model with the in-house reservoir simulation model for two of the simulation runs that were included in the training of the Smart Proxy Model. In Figure 4.22, the comparison is made for total field production (both oil and gas). Figures 4.23 and 4.24 show the comparison for randomly selected wells in two of the simulation runs that were included in the training of the Smart Proxy Model. It can be seen in all of these figures that the Smart Proxy Model can reproduce the results of the in-house simulator with high accuracy.

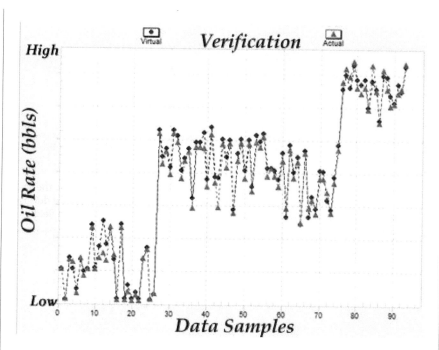

Figure 4.21 Performance of field "R" Smart Proxy Model on the 25% of randomly selected blind data from the spatio-temporal database.

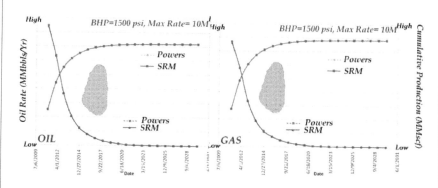

Figure 4.22 Comparing the accuracy of field "R" Smart Proxy Model with the in-house numerical reservoir simulator for a run that was included.

The trained Smart Proxy Model is validated against a complete blind run of the reservoir simulation model. For this purpose, a 10th run was made where the operational constraint was completely different (although within the range) from the runs that were made to build the spatio-temporal database for the training and matching of the Smart Proxy Model.

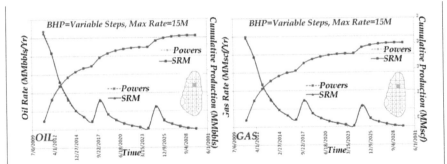

Figure 4.23 Comparing the accuracy of field "R" Smart Proxy Model with the in-house numerical reservoir simulator for a run that was included during the training; oil and gas production from one of the wells in the field.

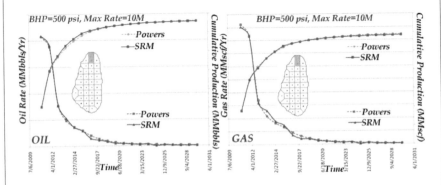

Figure 4.24 Comparing the accuracy of field "R" Smart Proxy Model with the in-house numerical reservoir simulator for a run that was included during the training; oil and gas production from one of the wells in the field.

The operational constraints of the blind run are BHP of 1,000 psi and maximum oil rate of 15,000 barrels. Figures 4.25–4.27 show the comparison of the Smart Proxy Model with the blind run of the in-house reservoir simulation model. In Figure 4.25, the comparison is made for oil and gas production for the entire field. In Figures 4.26 and 4.27, the comparison of two randomly selected wells in a blind run of the simulation model is shown.

It can be seen in all these figures that the SRM can reproduce the results of the in-house simulator with high accuracy, even with the operational constraints that it has never seen or been trained for. This shows the generalization and abstraction capability of the Smart Proxy Model. Using this capability sensitivity analysis, quantification of uncertainties and comprehensive exploration of the solution space for identification of optimum field development strategies become practical. All these analyses can be performed in record time.

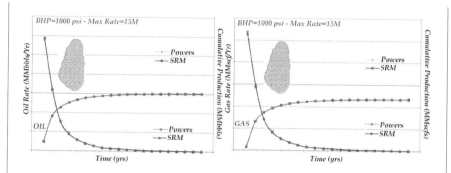

Figure 4.25 Comparing the accuracy of field "R" Smart Proxy Model with the in-house numerical reservoir simulator for a blind run; oil and gas production from all the wells in the field.

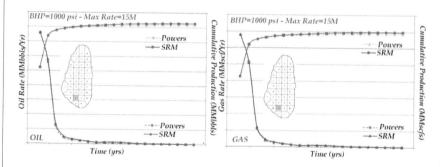

Figure 4.26 Comparing the accuracy of field "R" Smart Proxy Model with the in-house numerical reservoir simulator for a blind run; oil and gas production from one of the wells in the field.

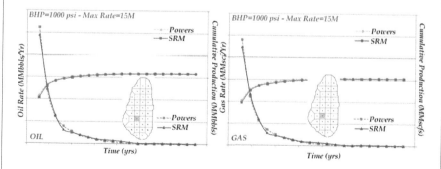

Figure 4.27 Comparing the accuracy of field "R" Smart Proxy Model with the in-house numerical reservoir simulator for a blind run; oil and gas production from one of the wells in the field.

As it was previously mentioned, due to limited number of wells that have been drilled in this field, considerable uncertainties exist in reservoir description as well as in the engineering data. One of the ways to address known uncertainties associated with the static model is to use Smart Proxy Model to perform sensitivity analysis.

The type of sensitivity analysis presented in this section can be used effectively during the history matching process of the NRS model. In the analyses presented here, type curves are developed to put in perspective "on" the sensitivity of production to various parameters at different locations in the reservoir. For example, for each individual well, these type curves can show how sensitive the production is to any given variable (i.e., permeability) at many predetermined locations in the reservoir. Therefore, it will not be necessary to modify the permeability of the entire reservoir using a "permeability multiplier" to achieve history match at a certain location (well) in the reservoir. Knowing which parameter to modify and how much, to achieve a history match, can help modelers make localized "small" modifications to key parameters and avoid large-scale modifications of the static model.

The results of the sensitivity analysis are presented in the form of a type curve. Type curves can be generated using the Smart Proxy Model for the individual wells, groups of wells, or for the entire field in seconds. Figures 4.28–4.30 are examples of several type curves that have been developed for field "R" using the Smart Proxy Model.

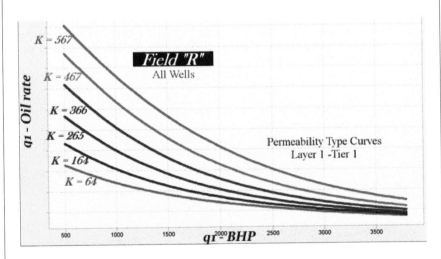

Figure 4.28 Sensitivity analysis of field "R" using the Smart Proxy Model. Sensitivity of oil rate to BHP for all the wells in the field (in general and on average) as a function of different values of permeability in the Tier One grid blocks (grid blocks closest to the wellbore).

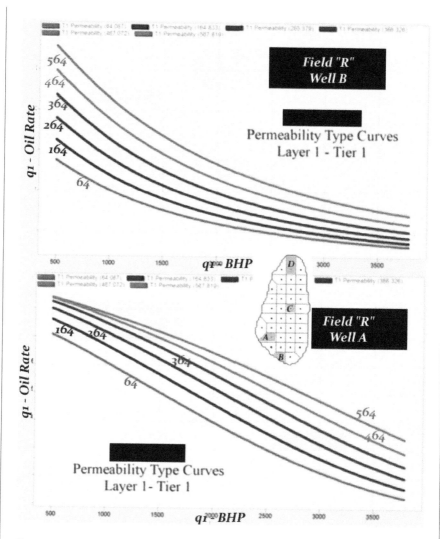

Figure 4.29 Sensitivity analysis of field "R" using the Smart Proxy Model. Sensitivity of the oil rate to BHP for Wells A and B in the field as a function of different values of permeability in the Tier One grid blocks (grid blocks closest to the wellbore).

Figure 4.28 shows the sensitivity of oil rate to BHP as a function of permeability in Tier One grid blocks in general for all the wells in field "R". This type curve is a general type curve that shows the overall impact of permeability on Tier One grid blocks. Although it is useful for comparison to other tier blocks in the reservoir, it cannot be a big help on a well-by-well basis. For the purposes

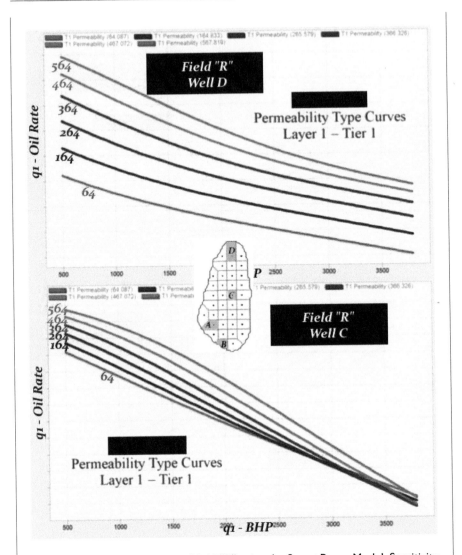

Figure 4.30 Sensitivity analysis of field "R" using the Smart Proxy Model. Sensitivity of oil rate to BHP for Wells C and D in the field as a function of different values of permeability in the Tier One grid blocks (grid blocks closest to the wellbore).

of achieving history match in any given well, one should use type curves that are shown in Figures 4.29 and 4.30.

Figures 4.29 and 4.30 demonstrate the sensitivity of oil rate to BHP as a function of permeability in Tier One grid blocks for four selected wells in different locations in the field. Figure 4.29 shows the sensitivity for Wells A and B that are located in the southern part of the reservoir while Figure 4.30 shows the sensitivity for Well C in the central part of the reservoir and Well D in the northern part of the reservoir. Please note that these four wells portray four different behaviors. This is due to all other parameters that are involved in flow of fluid in the reservoir. The heterogeneous nature of the reservoir dictates this nonuniform response to modification to any given parameters.

When a permeability multiplier is used to modify the permeability of the entire field, to achieve a history match for a given well, the sensitivity of different locations of the field to different parameter modifications is ignored. The difference in the response of the reservoir to modification of a given parameter is, in itself, an indication of the heterogeneity of the reservoir that needs to be honored and accounted for. We believe that using Smart Proxy Model in the way it is mentioned here is a better way of honoring overall reservoir heterogeneity and the work that has gone into developing the geological model in the first place.

As previously mentioned, the limited number of wells that have been drilled in this field result in considerable uncertainties in reservoir description. This is true with all green fields where the sample data from logs and core are so limited that analogy or general geological understanding of the field becomes the limits of our understanding of the reservoir heterogeneity. Although surface seismic may be available in many of the prolific green fields, deduction of reservoir parameters as needed for simulation and modeling is a stretch, and many scientists are still working on feasible solutions.

It is a universally agreed issue that uncertainties associated with the reservoir characteristics, and therefore the static model that is used for the analysis of green fields, pose a serious challenge to comprehensive modeling of these fields. Thus, uncertainty analysis in such cases becomes an important step that needs to be taken into account. On the other hand, as the number of grid blocks required to build a geologic model for the reservoir simulation models increases, so does the amount of time required for a single run.

As previously mentioned, in the case of field "R", 1.4 million active cells are used to model this reservoir. Even with massively parallel, multi-cluster computational resources, it takes about 12 minutes to make a single run. This means that a reasonably simple uncertainty analysis that requires only 1,000 simulation runs will take more than eight days of computational time and the slightest mistake or modification that would require the runs to be repeated will take another eight days. A more comprehensive uncertainty analysis that would require extensive sampling of the geological model with about 10,000 runs would take about three months, even with such massive resources. On the other hand, an uncertainty analysis with a moderate 1,000 runs will only take a few seconds with the Smart Proxy Model. Given the accuracy of the Smart

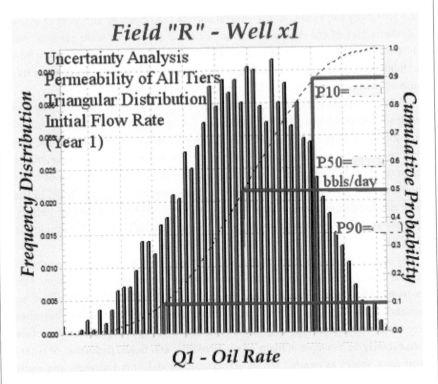

Figure 4.31 Uncertainty analysis of field "R" using the Smart Proxy Model. Uncertainty associated with permeability of the formation (all Tiers) for Well x1.

Proxy Model that has already been demonstrated, this is a significant advancement in the analytical power of the existing reservoir simulation models.

Figures 4.31–4.33 show the uncertainty analysis performed using field "R" Smart Proxy Model for three individual wells in this field. Each of these analyses was performed using 5,000 Smart Proxy Model runs and they took less than five seconds each. In these analyses, the total oil production during the first year was calculated using field "R" Smart Proxy Model when all the permeability values in the field were modified, sampling from the frequency distribution functions that were developed using all the available values in the simulation model.

This represented a large range of permeability values. But the range in such analyses can be changed to represent any minimum and maximum values as well as any type of distribution function (i.e., triangular, uniform, Gaussian, etc.). Along with representing the probability distribution that is presented in Figures 4.31–4.33, P10, P50, and P90 for each of these wells are also identified.

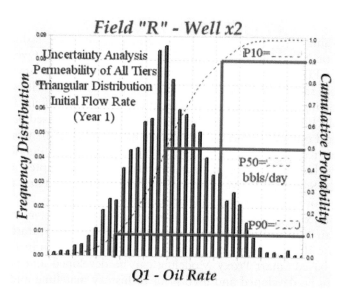

Figure 4.32 Uncertainty analysis of field "R" using the Smart Proxy Model. Uncertainty associated with permeability of the formation (all Tiers) for Well x2.

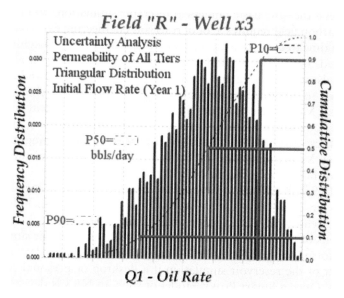

Figure 4.33 Uncertainty analysis of field "R" using the Smart Proxy Model. Uncertainty associated with permeability of the formation (all Tiers) for Well x3.

4.2 CELL-BASED SMART PROXY MODELING

Cell-based Smart Proxy Modeling is used almost in the same fashion when it is working on NRS and computational fluid dynamics (CFD). Overwhelming majority of CFD applications are in mechanical, chemical, and many other engineering areas rather than reservoir simulation. Nevertheless, cell-based Smart Proxy Modeling would be the same general way of using Artificial Intelligence (AI) and Machine Learning in order to build a correct AI-based proxy model for almost all types of numerical simulation models. The same would be true to apply this technology to build a Smart Proxy Model for numerical simulation models that are used for weather forecasting. In this chapter, an example of cell-based Smart Proxy Modeling in the NRS is explained and then recent case studies for both NRS and CFD are covered.

In the context of NRS, cell-based Smart Proxy Modeling refers to the development of an AI-based model representing the pressure and saturation at every time-step for all the existing cells (grid blocks) that have been included in the NRS. Cell-based Smart Proxy Modeling can be developed and used in two series. It can be developed and used prior to history matching and also after completion of history matching. Prior-history matching cell-based Smart Proxy Modeling is developed for the basic NRS that uses the original (basic) geological model provided to reservoir engineers by geologist and geo-physicists in order to be used for reservoir modeling. It is a well-known fact that "the basic numerical reservoir simulation" never creates an acceptable history match. Usually, using the "the basic numerical reservoir simulation" to history match the hydrocarbon field requires a lot of reservoir engineering expertise and takes a very long time to come to a correct and acceptable history matching result.

Cell-based Smart Proxy Modeling provides a tool that can enhance the quality of history matching and significantly reduce the time and efforts that are required to achieve history matching. Post-history matching cell-based Smart Proxy Modeling can be developed using the history matched NRS for uncertainty quantification and any type of optimization using the detailed characteristics of the NRS. The detailed process of developing a cell-based Smart Proxy Modeling for an NRS covered in this chapter would be exactly the same for both prior- and post-history matching.

When NRS is developed with reasonable size for a field that includes multiple reservoirs layers, a practical reservoir model ends up including several millions of cells (grid blocks). The large number of grid blocks of an NRS result in tens of hours for a single deployment of the reservoir simulation on a high performance computer (HPC), and probably multiple days for a single deployment of the reservoir simulation if a desktop or personal computer is being used. Once a Smart Proxy Model of such an NRS is developed (training, calibration, and validation), then a single deployment of the Smart Proxy Model that would provide over 95% accurate results of the reservoir simulation (reservoir pressure and saturation for every cell at every time-step) can be completed in minuets on desktops and even on laptops.

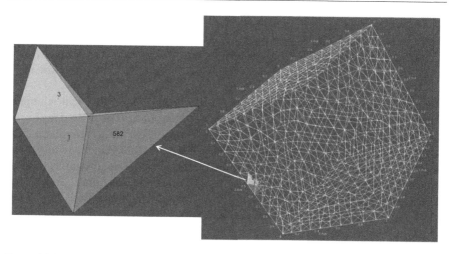

Figure 4.34 Numerical simulation developed by tetrahedron cells.

The example of the cell-based Smart Proxy Modeling and its detailed development that will be covered in the next section as "Case Study", is for a prior-history matching of NRS of CO_2 sequestration. The numerical simulation can include tetrahedron cells as shown in Figure 4.34 or cubic cells of different sizes as shown in Figure 4.35.

All explanations that will be presented in this chapter can be applied to any shape of the cells that will be used in the numerical simulation. Details of how cell-based Smart Proxy Model is developed will be covered in this section

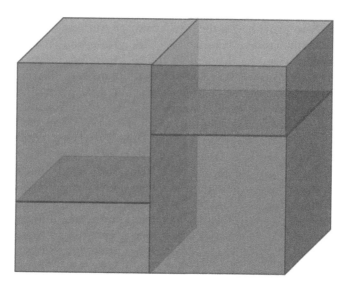

Figure 4.35 Numerical simulation developed by different sizes of cubic cells.

before providing a recent case study of a cell-based Smart Proxy Modeling in the next section. As mentioned before, cell-based Smart Proxy Model is all about the cells in the NRS. In order to be able to fully understand the development of the cell-based Smart Proxy Model, it is important to know how NRS is developed based on cells.

Since AI is developed through Machine "Learning" algorithms, it is important to note how machines would "learn"? Is it enough to provide data to the Machine Learning algorithms? Those who might believe that is the main definition of Machine Learning, most probably think that Machine Learning is the same as traditional Statistics. What some people seem to be missing is the difference between "Artificial Intelligence & Machine Learning" and "Traditional Statistics". What today is called "Artificial Intelligence & Machine Learning" was literally started in mid-1980s (about 35 years ago) while the "Traditional Statistics" was developed more than 120 years ago. How could these two things possibly be the same? While "Traditional Statistics" is all about providing only the data that is collected to its algorithms, the same is NOT TRUE about the engineering application of "Artificial Intelligence & Machine Learning". When it comes to the engineering problem solving (that is the main characteristics of this book), Machine "Learning" requires expert "Teaching". In other words, there are many differences between different types of "Learning". "Learning" without "Teaching" provides superficial results.

Here is an example. Imagine you are opting two engineering technical courses at a university. In one of the courses (Course "S"), the professor that is in charge of the course would walk in the first class and have the following communication with you: "Here is the book for this course, and here are the few technical papers about this course. Please learn all about the topics in this book and these technical papers and I will meet you again at the end of the semester to provide you the final exam for your grades". On the other hand, in the second course (Course "A"), the professor that is in charge of the course meets with you throughout the entire semester, every week. The professor gives lectures in every course meeting, answers all your questions, covers all the details about every topic in the course, solves multiple problems for you, provides you with several homework, quizzes, and exams throughout the semester, and covers the solutions of all the problems in the homework, quizzes, and exams before the end of the semester when she/he would give you the final exam.

Here is a couple of simple questions: (1) Can you conclude that both professors of these two courses that you took, were domain experts in the topic of the engineering technical courses? (2) Have you learned the same way from the two engineering technical courses that you took at the university: Course "S" and Course "A"? If you think the answer is "yes" to these two questions, then you are correct to assume that "Traditional Statistics" is the same as "Artificial Intelligence & Machine Learning" since you were exposed to the topic by domain experts and have learned exactly the same way in both courses.

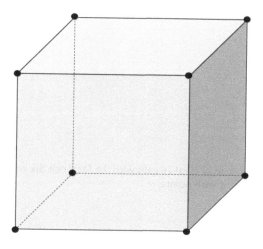

Figure 4.36 Simple cubic cell used in most NRS models.

However, this is NOT TRUE. Here are the two main characteristics: (a) Actual "Learning" requires actual "Teaching" rather than just providing information, (b) actual "Teaching" requires domain expertise. Hopefully this simple example can remind you that "Learning" through "Teaching" requires domain expertise. What is being covered in this section about cell-based Smart Proxy Modeling is all about "Learning" through "Teaching". For the simplification purposes, let us start explaining the cell-based Smart Proxy Modeling process through a simple cubic cell, shown in Figure 4.36.

Since numerical simulation is the simplification of the solution of the complex mathematical equations that is being used to model the intricate physical phenomena through digitalization in space and time, it starts with a cell. The size (volume) of the cell (several millions of which is required to model the physical system being modeled) must be small enough to make all characteristics of the complex mathematical equations to be linear and homogeneous, so the mathematical equations can be solved easily and simply. In Figure 4.36, it is shown that the simple cubic cell includes six planes, twelve lines, and eight points. This shows that each of cells in the several millions of cells in the NRS is connected and communicates with 26 (6 + 12 + 8 = 26) surrounding cells.

The 26 surrounding cells communicate with the focal cell through three different tiers. Tier 1 offset cells include six cells. Tier 1 offset cells mostly communicate with the focal cell through planes. This is shown in Figures 4.37 and 4.38 (on the left, identified as plane communication). Tier 2 offset cells include twelve cells. Tier 2 offset cells have the second level of communication with the focal cell through lines. This is shown in Figures 4.37 and 4.38 (on the middle, identified as line communication). Tier 2 offset cells communicate with the focal cell through two steps of the plane communication. Their plane communication is with Tier 1 offset cells.

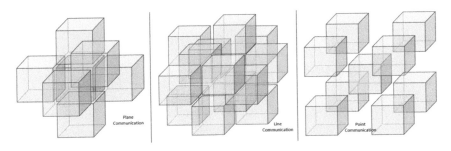

Figure 4.37 Communication of total 26 cells with the focal cell. Six with planes, twelve with lines, and eight with points.

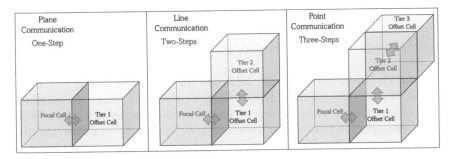

Figure 4.38 Communication of Tiers 1, 2, and 3 offset cells with the focal cell.

Tier 3 offset cells include eight cells. Tier 3 offset cells have the third level of communication with the focal cell through points. This is also shown in Figures 4.37 and 4.38 (on the right, identified as point communication). Tier 3 offset cells communicate with the focal cell through three steps of the plane communication. Their plane communication is with Tier 2 offset cells. Based on the diffusivity equation that is a parabolic partial differential equation, pressure and saturation in each single (focal) cell change during the NRS at each time-step. Since each cell communicates with 26 other cells as shown in Figures 4.37 and 4.38, modification of pressure and saturation in each cell is a function of the modification of pressure and saturation in the 26 neighboring cells.

Furthermore, it must be noted that other objectives that would play a role and contribute to the pressure and saturation modification in each cell in the NRS would be a function of several other characteristics. These characteristics include:

a. The location of the cell in terms of its distance from the reservoir boundaries,
b. The location of the cell in terms of its distance from any injection and production wells,

c. Reservoir characteristics such as porosity and permeability for each of the cells that is between the injection and production wells and the focal cell, and

d. The way the injection and production wells have been completed.

This means that all the information and knowledge about how an NRS is developed and solved, must be used in order to teach the Machine Learning algorithm about what is happening based on the data that would be used to build the AI-based Smart Proxy Model.

As mentioned earlier, the idea behind the development of the cell-based Smart Proxy Model is to teach the Machine Learning algorithm how NRS was developed and how the pressure and saturation at every time-step in a given cell is changed. It is a well-known fact that the only way to teach the Machine Learning algorithm is through providing data. Figures 4.39–4.43 provide examples of how such teaching through data can be performed for training the Machine Learning algorithm. Since it is easier to show details in two-dimensional (2-D) figures rather than three-dimensional (3-D) figures, Figures 4.39–4.43 show the details of how information (data) for each focal cell should be generated for teaching and training the Machine Learning algorithm.

Figure 4.39 shows the location of a focal cell and its offset cells. Figure 4.40 shows the distances of the focal cell to reservoir boundaries, and Figure 4.41 shows the distances of the focal cell to the existing production and injection wells. It is important to note that distances of each cell to reservoir boundaries and all the existing production and injection wells have an impact on the reservoir pressure and fluid saturation of the well at each time-step. It is also

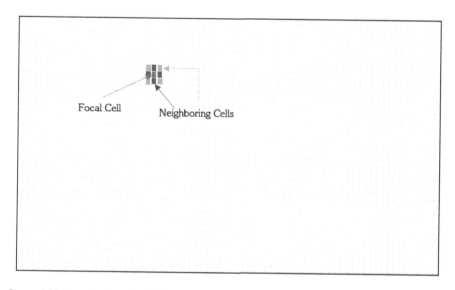

Figure 4.39 Focal cell and neighboring cells in a 2-D reservoir model.

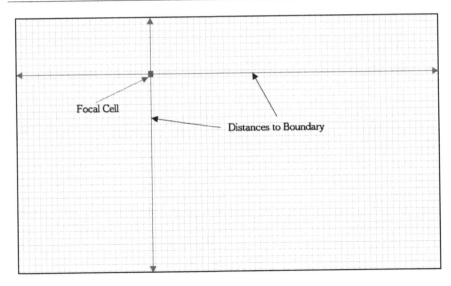

Figure 4.40 Distances of the focal cell from the boundaries of the reservoir in 2-D reservoir model.

important to note that the number of injection and production wells that would exist at every time-step plays an important role in inputting data to the Smart Proxy Model development.

Given the fact that the operational conditions, completion details, and measured amounts of fluids for each injection and production well play an

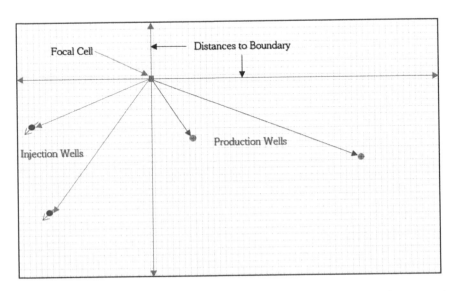

Figure 4.41 Distances of the focal cell from the injector and producer wells in the reservoir in 2-D reservoir model.

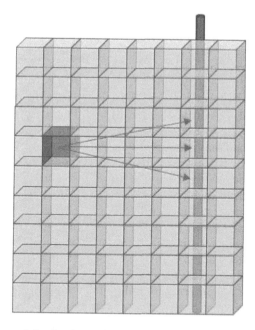

Figure 4.42 Distances of the focal cell from the completed cells (same layer and a layer below and above) of an injection or production well in 2-D reservoir model.

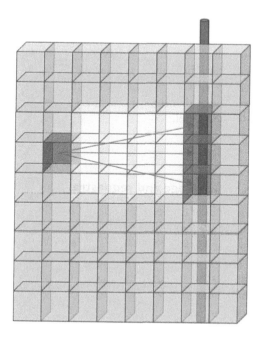

Figure 4.43 The cells between focal cell and the completed cells (same layer and a layer below and above) of an injection or production well in 2-D reservoir model.

important role in the changes of reservoir pressure and fluid saturation for every cell in the NRS model, it is important to include such data as input for modeling the AI-based reservoir pressure and fluid saturation of the Smart Proxy Model for every time-step.

Figures 4.42 and 4.43 show 2-D example parts of an NRS model with an existing well and a focal cell. Figure 4.42 shows the distance of the focal cell from the completed cells of the existing well and the distances to a completed cell above and one below the same layer where the focal cell exists. Since in NRS, each cell usually has permeability values in x, y, and z directions, therefore the completed cell above and below the completed cell in the same layer of the focal cell would also play a role in reservoir pressure and fluid saturation of the focal cell in each time-step.

Thus, it is a good idea to include these cell distances from the focal cell in the input of the Smart Proxy Model development. Figure 4.43 shows the cells whose reservoir characteristics play a role in how the reservoir pressure and fluid saturation are modified in a specific time-step. Therefore, it would be a good idea to include the average of the reservoir characteristics of such several cells in the input of the Smart Proxy Model development.

What has been covered in this section of this chapter provides examples of how to perform feature generation when the Smart Proxy Model development is being started. It should also be noted that it does not mean that these are the only required feature generations, or all of them must be included in the input characteristics of the Smart Proxy Model development. During training, calibration, and validation of the Smart Proxy Model development, the results will provide you with the information indicating that you are moving in the right direction for the Smart Proxy Model development.

In the next two cased studies where special Smart Proxy Modeling development for an NRS and a CFD is presented, it can be seen what actual inputs from each of these numerical simulation models were used in order to generate a good Smart Proxy Model.

Case study: CO_2 storage and sequestration in saline aquifer

*Co-author of this specific section of this chapter is
Maher Al Aboody from West Virginia University Laboratory for
Engineering Application of Data Science (WVU-LEADS).*

In the next several decades, storage and sequestration of carbon dioxide (CO_2) into geological formations will emerge as the major solution to address the issue of climate change. Currently, overwhelming majority of carbon storage and sequestration is performed in saline aquifers. From a reservoir engineering point of view, carbon storage and sequestration in saline aquifers is a reservoir modeling of a green field. Green field is defined as a reservoir (field) in which minimum measured reservoir characteristics are available. This is due to the fact that prior to carbon storage and sequestration, no other wells have been drilled in the specific saline aquifer. All realistic measured reservoir characteristics can only be available through well logs and cores that have been performed through drilled wells in a given reservoir (field).

NUMERICAL RESERVOIR SIMULATION MODEL

The reservoir model was obtained from a history match model created at WVU-LEADS previously [Haghighat and Mohaghegh, 2015] on the Citronelle field, a saline aquifer reservoir located in Mobile County (Alabama, US). The structure of the numerical reservoir model includes 125 cells in x direction, 125 cells in y direction, and 65 layers in z direction. The total number of cells in this NRS is more than one million: $125 \times 125 \times 65 = 1,015,625$.

The reservoir includes five different layers as shown in Figure 4.44. There are four simulation shale layers on the top, six simulation shale layers in the middle, and four simulation shale layers in the bottom. These 14 simulation shale layers act mainly as blocking layers that would not allow the injected CO_2 to move above or below of total 51 simulation sandstone layers (24 simulation sandstone layers above and 27 simulation sandstone layers below) that would include the injected CO_2. Therefore, total 65 simulation layers (five formation reservoir layers) include 14 simulation shale layers (three formation shale layers) and 51 simulation sandstone layers (two formation sandstone layers).

The sandstone layer on the top includes 24 simulation layers, and the sandstone layer in the bottom includes 27 simulation layers. Actual porosity and permeability measurements from the saline aquifer of the Citronelle field, as

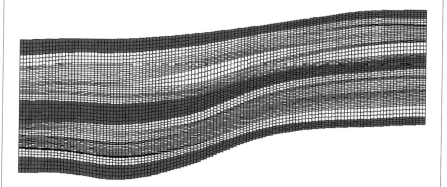

Figure 4.44 The NRS includes five layers (three shale layers and two sand layers). As shown in this figure, the sizes of the cubic cells are not all the same.

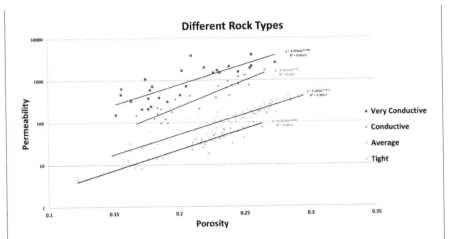

Figure 4.45 Porosity-permeability characteristics of the rock types used in the NRS.

shown in Figure 4.45, were used to generate four rock types. These four rock types are used in this NRS. The top 24 simulation sandstone layers are divided into two rock types (rock type C4, very conductive [12 simulation sandstone layers] and rock type C3, conductive [12 simulation sandstone layers]). The bottom 27 simulation layers are also divided into two more rock types (rock type C2, average [14 simulation sandstone layers] and rock type C1, tight [13 simulation sandstone layers]). Distribution of 65 simulation layers into seven different formation layers is shown in Figure 4.46.

Injection of CO_2 into this saline aquifer is done using four injection wells. These four CO_2 injection wells are shown in 3-D version of the numerical simulation field in Figure 4.47. Each of these four injection wells are completed in all the 51 simulation sandstone layers in the NRS.

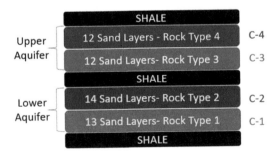

Figure 4.46 Fifty-one simulation layers of sand are divided into four different rock types.

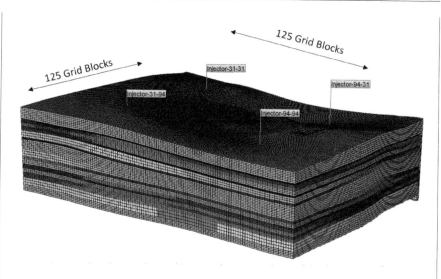

Figure 4.47 3-D shape of the NRS, including four injection wells.

As shown in Figure 4.48, the NRS is to be performed for 300 years from January 2020 to January 2320. The four injection wells have been injecting CO_2 for 30 years and then the injection is stopped. The NRS runs have provided reservoir pressure and CO_2 saturations throughout the 30 years of injection and would continue the same during the next 270 years after the injection is stopped.

Furthermore, given the fact that this is the NRS of a green field, 20 different geological realizations of the entire field (specifically the 51 sand layers) were generated. As shown in Figure 4.49, only 80% of these geological realizations (16 out of 20) are used for the development of the Smart Proxy Model. The other 20% of these geological realizations (4 out of 20) are used as blind validation simulation runs in order to realistically check and confirm the strong quality of the predictions that will be made by the developed Smart Proxy Model.

The development of the Smart Proxy Model includes generation of comprehensive data set from the simulation runs that are used. Since 16 geological

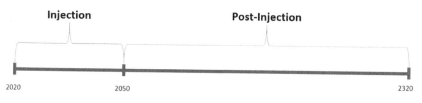

Figure 4.48 The NRS used for CO_2 sequestration includes 300 years – 30 years of injection and 270 years of post-injection.

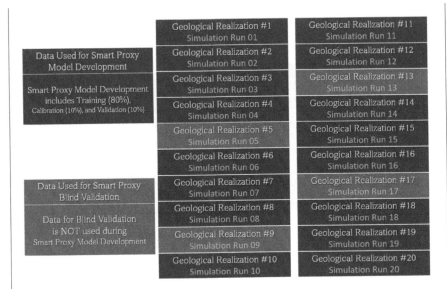

Figure 4.49 List of the 20 geological realizations.

realizations are used for the development of the Smart Proxy Model, they are used as the geological model to make 16 NRS runs, each for 300 years. Upon completion of the NRS runs, the comprehensive data sets are generated from each of the 16 simulation runs. The data used for the development of the Smart Proxy Model will be used for training, calibration, and validation. This complete data set is randomly divided into three categories such as 80% for training, 10% for calibration, and another 10% for validation.

GEOLOGICAL REALIZATIONS

The four randomly selected blind validation geological realizations that are shown in Figure 4.49 are *not* used in any shape or form during the development (training, calibration, and validation) of the Smart Proxy Model. These blind validation geological realizations are used to make a decision on the realistic quality of the Smart Proxy Model that is developed. The main reason is that the Smart Proxy Model will be used for uncertainty quantification and any shape or form of optimization that requires millions of NRS runs. When the Smart Proxy Model can be proven to perform correctly with minimal error using completely new geological realizations that were not used to develop it, then its use will make sense and can be trusted.

Figures 4.50–4.53 show 20 geological realizations. These four figures provide porosity and permeability distributions on two of the layers (Layer 5 – very conductive rock type and Layer 50 – tight rock type) as examples. Similar

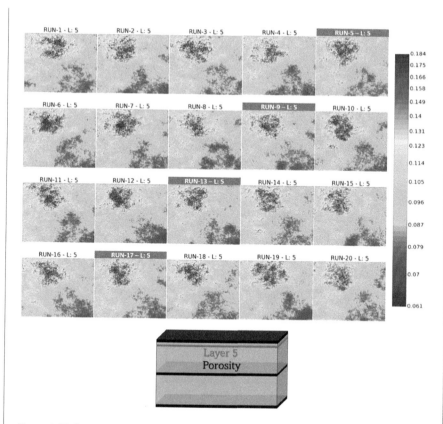

Figure 4.50 Porosity distribution of layer 5 (very conductive rock type – shown in blue line) for 20 geological realizations, 16 of which are used for the Smart Proxy Model development and 4 for blind validation.

distributions have been done on all the 51 simulation layers in the 20 geological realizations. In these figures, geological realization numbers 5, 9, 13, and 17 are also identified that are used as blind validations.

While the images in Figures 4.50–4.53 show porosity and permeability distributions for 4 out of 51 layer of the geological realizations, it is important to note that Smart Proxy Modeling is not developed as a function of such images. It is a fact that some individuals that are trying to build AI-based models for numerical simulations, use convolutional neural networks (CNN), instead of engineering application of AI and Machine Learning. Such individuals are faculty in academia, engineers in National Labs, and data scientists in service companies, start-up companies, and even operating companies. Using CNNs means that they are trying to perform image recognition rather than using engineering application of AI and Machine Learning. This usually exposes their actual understanding of AI and Machine Learning.

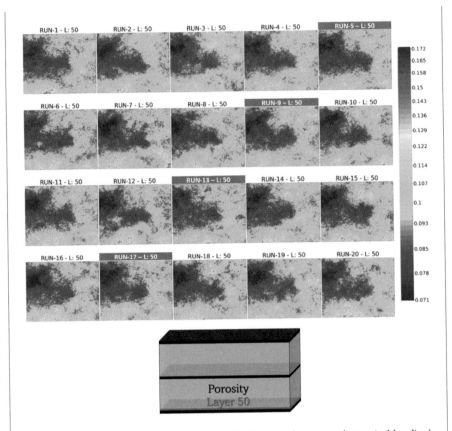

Figure 4.51 Porosity distribution of layer 50 (tight rock type – shown in blue line) for 20 geological realizations, 16 of which are used for the Smart Proxy Model development and 4 for blind validation.

Most of the time, they use images similar to what is shown in Figures 4.50–4.54 to perform CNNs. When such an approach is performed in the context of reservoir engineering, it clearly identifies one of the following two possibilities about those that perform such approaches:

a. They are AI experts with very little or no understanding of reservoir engineering, or
b. They are reservoir engineering experts with very little or no understanding of AI.

The fact is that reservoir engineering/modeling is not an image recognition approach to problem solving. The 2-D or 3-D images of each reservoir layer that

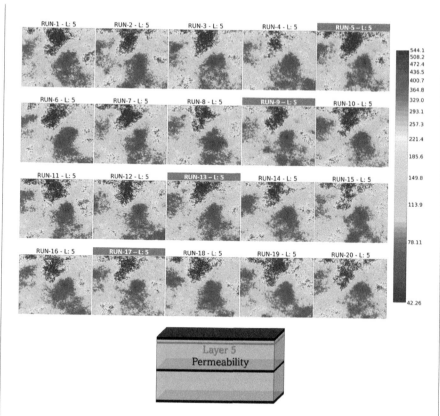

Figure 4.52 Permeability distribution of layer 5 (very conductive rock type – shown in blue line) for 20 geological realizations, 16 of which are used for the Smart Proxy Model development and 4 for blind validation.

are presented and looked at only provide information about the characteristics of the reservoir heterogeneity, which contributes to the reservoir pressure and saturation distribution as a function of space and time. The main reason that such images are important in Smart Proxy Modeling is that these are not used as an image for AI-based model development or training neural networks, rather these images provide information about the impact of heterogeneous reservoir characteristics that impact and contribute to pressure and saturation distribution as a function of space and time.

Pressure and saturation distribution throughout the entire reservoir as a function of space and time is also a function of injection wells including their locations, their completions, their flowing bottom hole pressure constraints, as well as their injection rates. Since completions, bottom hole pressure, and

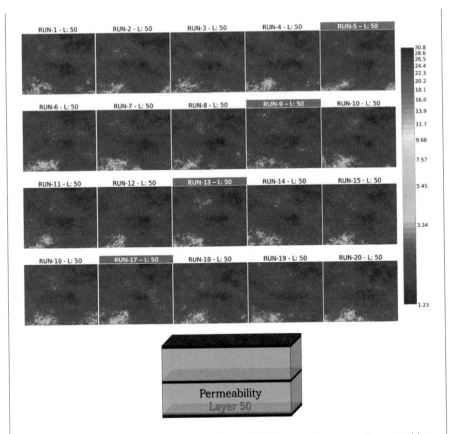

Figure 4.53 Permeability distribution of layer 50 (tight rock type – shown in blue line) for 20 geological realizations, 16 of which are used for the Smart Proxy Model development and 4 for blind validation.

injection rates may change on a daily, weekly, or monthly basis, CNN will not be able to provide information that can be used for development and/or blind validation of AI-based model of the reservoir simulation.

Instead, such images provide reservoir engineers an opportunity to understand the reservoir characteristics that are going to be used to calculate the reservoir pressure and saturation distribution throughout the entire reservoir. What Smart Proxy Modeling does, is using all the spatial and temporal characteristics of fluid flow in the porous media, in order to teach the Machine Learning algorithms how to generate reservoir pressure and CO_2 saturation for every single cell in the reservoir model at every time-step. This is the reason that this section is called cell-based Smart Proxy Modeling. Detailed data and information in space and time about each single cell are used in order to develop AI-based simulation model.

INJECTION WELLS

As shown in Figure 4.47, there are four CO_2 injection wells in this NRS model. These four CO_2 injection wells are named "Injector 31-31", "Injector 94-31", "Injector 31-94", and "Injector 94-94" that represent the "i, j" cell number where the CO_2 injection wells are located.

During the 20 simulation runs that have been generated (a 300-year simulation run for each geological realization), each of the four injection wells includes a specific CO_2 injection rates and BHP constraint. The BHP constraint was 5,500 psi, and the CO_2 injection rate constraint was 2.84 million metric tons per year.

Figures 4.54 and 4.55 show the details of BHP and CO_2 injection rate for every single injection well for the 20 NRS runs, each using a different geological realization. Their examples are shown in Figures 4.50–4.53. It is interesting and

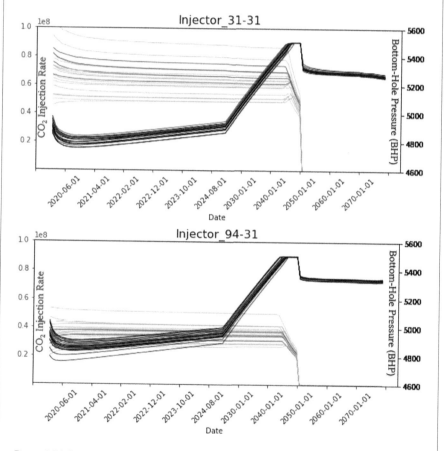

Figure 4.54 Details of BHPs and injection rates for CO_2 injection wells "31-31" and "94-31" for 20 simulation runs, each related to a separate geological realization.

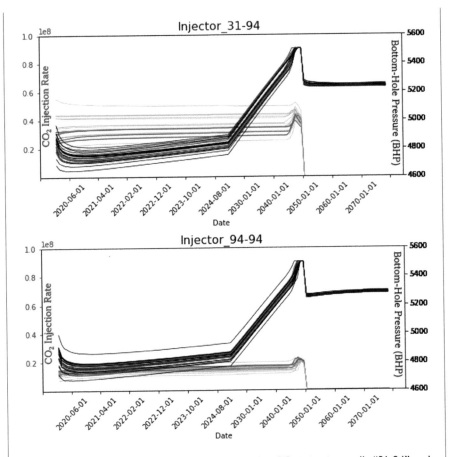

Figure 4.55 Details of BHPs and injection rates for CO_2 injection wells "31-94" and "94-94" for 20 simulation runs, each related to a separate geological realization.

also very important to note that historical data of both BHP and CO_2 injection rate for each well is different in each of the 20 simulation runs.

SMART PROXY INPUTS AND HYPERPARAMETERS

As mentioned earlier, the Smart Proxy Model for this CO_2 sequestration NRS model is based on each cell (grid block) that is included in the NRS. Therefore, all the input data used for the development of the Smart Proxy Model refer to the detailed characteristics of each cell. Given the fact that NRS modeling is a cell-based model, any spatial and/or temporal characteristic that is directly or indirectly used to calculate the reservoir pressure and CO_2 saturation for each cell at every time-step must be brought into the development of the AI-based Smart Proxy Model. These are the data (information) that would help to teach reservoir engineering details to the Machine Learning algorithms that are used to develop the Smart Proxy Model.

Following are the seven categories of the input data for the training and development of the Smart Proxy Model:

Cell location
Initial dynamic data for each cell
Formation properties for each cell
Distances to reservoir boundaries for each cell
Details of "Tier One" (plane communication) offset cells for each Focal cell
Details of "Tier Two" (line communication) offset cells for each Focal cell
Injection data

Let's discuss and provide details about each of the seven categories of data that is used as the input of the Smart Proxy Model:

CELL LOCATION

The location of each cell that is referred to as the focal cell includes i, j, and k of the cell in the context of the total number of cells in the NRS. Furthermore, X, and Y values of the center of the focal cell from any given source of measurements will be included as the data for cell location. The total data of cell locations for a focal cell would be "5 Inputs".

INITIAL DYNAMIC DATA FOR EACH CELL

Given the fact that the dynamic data that is being calculated by the NRS and also will be the output of the Smart Proxy Model are reservoir pressure and CO_2 saturation, then their initial value would also be used as the input. However, since the initial CO_2 saturation in this case is zero for all cells, it does not require to be an input. The only initial data that would be required is reservoir pressure, which is a function of the reservoir depth at each focal cell. Therefore, the total data of initial dynamic data would be "1 Input".

FORMATION PROPERTIES FOR EACH CELL

The formation properties that are used in NRS for each cell are porosity, permeability, thickness, top, bottom, and center of the focal cell. The total data of formation properties for a focal cell would be "6 Inputs".

DISTANCES TO RESERVOIR BOUNDARIES FOR EACH CELL

Another important characteristic of each focal cell is its distances from East, West, North, and South of the reservoir boundary as well as it distances from

"Top Seal" (the shale formation) and the "Bottom Seal". The total data of distances to reservoir boundaries for a focal cell would be "6 Inputs".

DETAILS OF "TIER ONE" (PLANE COMMUNICATION)
OFFSET CELLS FOR EACH FOCAL CELL

As shown in Figure 4.37 (the left figure), there are six cells that are communicating with each focal cell through plane. These are known as Tier One offset cells. For each of these cells, characteristics such as porosity, permeability, thickness, top, and initial reservoir pressure would be used as input. Therefore, five inputs will be used for six Tier One offset cells. The total data of Tier One offset cells for a focal cell would be "30 Inputs".

DETAILS OF "TIER TWO" (LINE COMMUNICATION)
OFFSET CELLS FOR EACH FOCAL CELL

As shown in Figure 4.37 (the middle figure), there are 12 cells that are communicating with each focal cell through lines. These are known as Tier Two offset cells. For each of these cells, characteristics such as porosity, permeability, thickness, top, and initial reservoir pressure would be used as input. Therefore, five inputs will be used for twelve Tier Two offset cells. The total data of Tier Two offset cells for a focal cell would be "60 Inputs".

INJECTION DATA

Since there are four injection wells in this NRS model, all the input data that is mentioned here must include all these four wells. However, from a reservoir engineering point of view, the characteristics of the focal cell must be a function of its distance to the injection well. In other words, these four injection wells will NOT be referenced as Injection Well numbers 1, 2, 3, and 4 for all the cells in the model. Instead, each of the injection well will be referred to as number 1 Closest Injection Well, number 2 Closest Injection Well, number 3 Closest Injection Well, and number 4 Closest Injection Well. This means that each injection well will have a different number of closest injection well based on the location of the focal cell.

The inputs used from the closest injection wells are distance, well-based injection BHP, well-based injection rate, cell-based injection BHP, and ell-based injection rate for four wells. The total data of injection data would be "20 Inputs".

Therefore, the total number of inputs used for this particular Smart Proxy Model is 128 as shown below:

$$5 + 1 + 6 + 6 + 30 + 60 + 20 = 128$$

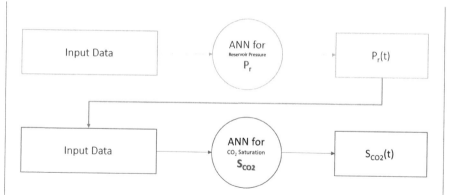

Figure 4.56 Smart Proxy Model for CO_2 sequestration in saline aquifer includes two Artificial Neural Networks, one for reservoir pressure and one for CO_2 saturation.

Once the input parameters have been identified, the process for developing the Smart Proxy Model has to do with the two Artificial Neural Networks. One for training, calibration, and validation of the reservoir pressure for every cell in the NRS, and one for training, calibration, and validation of CO_2 saturation for every cell in the NRS. This process would help the development of the CO_2 saturation neural network since the generated reservoir pressure by the first neural network could be used as an input to the second (CO_2 saturation) neural network, as shown in Figure 4.56.

Given the fact that using fluid flow in porous media in NRS modeling makes sure that reservoir pressure and CO_2 saturation are related to one another in space and time, such interaction between the two Artificial Neural Networks will result in effective characteristics of the Smart Proxy Model development.

Once the input data shows that the target of the neural network during training and calibration provides reasonable results, then it would make sense to start optimizing the hyperparameters of the neural network that is being used for training, calibration, and validation. It is important to note that in engineering application of AI and Machine Learning development, generation and selection of input data is far more important for successful training and calibration of the neural network than the hyperparameter optimization. Figure 4.57 shows the list of hyperparameters that have been used in the neural networks in this Smart Proxy Model.

RESULTS OF THE CO_2 STORAGE SMART PROXY MODEL

First, the Smart Proxy Model was generated in several specific time-steps throughout the 300 years of simulation runs. As mentioned earlier, the 300 years of the simulation runs included 30 years of CO_2 injection through 4 injection

Neural Network Hyper Parameters	
Parameters	Values
Max Number Epoch	10,000
Number Hidden Layer	1
Neurons / Hidden Layer	1000
Batch Size	20,000
Hidden Layer Activation Function	Relu
Output Layer Activation Function	Tanh
Initial Learning Rate	0.001
Learning Rate Reduce patience	10 epoch
Reduce Factor	0.25
Minimum Learning Rate	0.00001 (1.0E-05)
Early stop patience	50 epoch
Early stop min tolerance	0.000001 (1.0E-06)

Figure 4.57 List of the neural network hyperparameters used in Smart Proxy Model.

wells and 270 years of post-injection. The idea is to learn the distribution and change in reservoir pressure and CO_2 saturation for every cell in the NRS during and after CO_2 injection.

Since the main objective of this project was to develop a dynamic Smart Proxy Model for CO_2 injection and storage in the saline aquifer, performing modeling for the specific time-steps prior to the dynamic modeling would provide the required information for the development of the dynamic Smart Proxy Model. It is important to note what is the definition and characteristic of dynamic Smart Proxy Model.

For a given saline aquifer that is the goal of CO_2 storage and sequestration, a green field numerical reservoir simulation is developed for the purposes to history match the real reservoir pressure and CO_2 plum distribution as it is measured in multiple locations of the saline aquifer as a function of time and space, optimize CO_2 injection, quantify the uncertainties associated with the model that is used for decision-making, and so on. Given the fact that usually such NRS includes millions of cells (grid blocks), a single deployment of the simulator, even on an HPC, takes a long period of time.

This multiple hour's requirement to make a single deployment of the NRS on an HPC makes any realistic type of history matching, identification, optimization, and uncertainty quantifications that are required for decision-making a very long and, in some cases, an impossible process to be realistically accomplished.

When a dynamic Smart Proxy Model is successfully developed, it generates highly accurate results of the detailed target (output) of the numerical simulator within a few minutes, and not only on an HPC but also on a desktop or a laptop. To make a run (deployment) of the dynamic Smart Proxy Model, the only thing that is needed would be a geological realization, not the developed numerical reservoir simulator.

Once a new geological realization of the given saline aquifer is provided, then the dynamic Smart Proxy Model can generate monthly or annual distribution of reservoir pressure and CO_2 saturation for every cell in the entire saline aquifer in a few minutes on a desktop or laptop. This would make history matching, identification, optimization, and uncertainty quantification an easy-to-accomplish objective. Unlike the traditional statistical techniques that used to be done for the development of proxy models (some commercial software applications, some researchers, and developers still use traditional statistics), the dynamic Smart Proxy Model does not require hundreds of geological realizations and simulation runs for its development.

This new AI-based technology can be developed using only a small number of geological realizations and/or only a small number of simulations deployments. The questions may be asked, "why does dynamic Smart Proxy Model require a small number of geological realizations and numerical simulations runs while the traditional statistical techniques require several hundreds of geological realizations and numerical simulations runs to develop proxy models?

Furthermore, the traditional statistical techniques to develop proxy models never achieve more than 70 to 80% accuracy of results of the NRS and also never achieve the results that represent the entire details of the NRS even when several hundreds of geological realizations and numerical simulations runs are used. Then, how is it that Smart Proxy Model can achieve more than 95% accuracy and capable of replicating the entire details of the NRS in space and time"?

The answers to these questions are:

a. Smart Proxy Model does not use pictures of reservoir pressure and CO_2 plum distributions in order to replicate the NRS.
b. Smart Proxy Model does not use only certain portions of the input and output of the NRS.
c. Smart Proxy Model uses detailed information of every cell (grid block) in each of the geological realizations and numerical simulations runs.
d. Since millions of cells (grid blocks) are involved in each of the geological realizations and numerical simulations runs, about ten geological realizations and numerical simulations runs will be generating tens of millions of

data that will be used to train the Machine Learning algorithms for the use of AI to develop the Smart Proxy Model.
e. The development of Smart Proxy Model has as much to do with domain expertise in numerical simulation and its related engineering technology as it has to do with AI and Machine Learning.

As shown earlier, the dynamic Smart Proxy Model that is presented in this chapter has been developed only by 16 geological realizations and simulation runs. In the continuation of this chapter, results of Smart Proxy Modeling of this NRS of the CO_2 storage in saline aquifer will be presented.

Four geological realizations (geological realization and their NRS runs numbers 5, 9, 13, and 17) that were originally generated were randomly selected not to be used in any shape or form during the development (training, calibration, and validation) of the Smart Proxy Modeling. These four geological realizations and their NRS runs are referred to as blind validation geological realizations and their NRS runs.

The main reason of creating and testing "Blind Validations" is that once the Smart Proxy Model is developed, it must be trusted that it is capable of generating the results of NRS (reservoir pressure and CO_2 saturation for every cell at any given time-step) with high accuracy in a short period of time, so that it can be used for history matching purposes, identification, optimization, and uncertainty quantification of CO_2 storage and sequestration in a specific saline aquifer. In the rest of this chapter, details of reservoir pressure and CO_2 saturation generation by the Smart Proxy Model as a function of time and space will be presented in two and three dimensions.

Please note that the dynamic Smart Proxy Model is developed based on generating the reservoir pressure and CO_2 saturation for every cell (grid block) in space and time for every geological realization, and the figures shown in this chapter represent the same thing in different fashions. In other words, no matter the figures are in 2-D or 3-D cased and also in single layer for multiple time-steps or for a single time-step for multiple layers, they actually represent the results of the dynamic Smart Proxy Model for every cell in the NRS. Characteristics of results generated by the dynamic Smart Proxy Model are the same as the characteristics of results generated by NRS.

RESERVOIR PRESSURE RESULTS OF THE CO_2 STORAGE SMART PROXY MODEL

Figures 4.58–4.61 show examples of comparing the results of NRS with the results generated by Smart Proxy Model for a specific time-step of the reservoir pressure distribution in saline aquifer. As mentioned earlier, the Smart Proxy

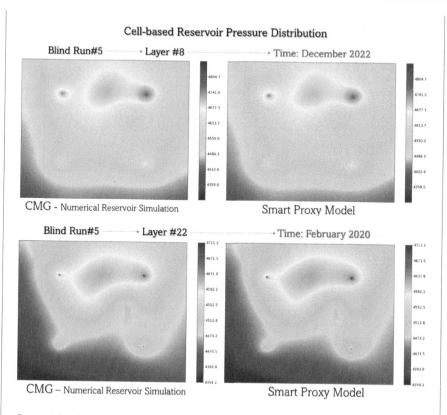

Figure 4.58 Comparing the results of NRS with the results generated by Smart Proxy Model for a specific time-step of the reservoir pressure distribution in saline aquifer. Blind validation (geological realization) run number 5; top figure: Layer 8 in December 2022, and run number 5; bottom figure: Layer 22 in February 2020.

Model generates the results (reservoir pressure and CO_2 saturation) for every single cell in the NRS at every time-step. These four figures show examples of certain layers and time-step for all four blind validation simulation runs. Figure 4.58 shows the reservoir pressure distribution of blind validation geological realization and its NRS run number 5 for layer 8 in time-step December 2022, and layer 22 in time-step February 2020.

Figure 4.59 shows the reservoir pressure distribution of blind validation geological realization and its NRS run number 9 for layer 26 in time-step January 2320, and layer 8 in time-step January 2035. Figure 4.60 shows the reservoir pressure distribution of blind validation geological realization and its NRS run

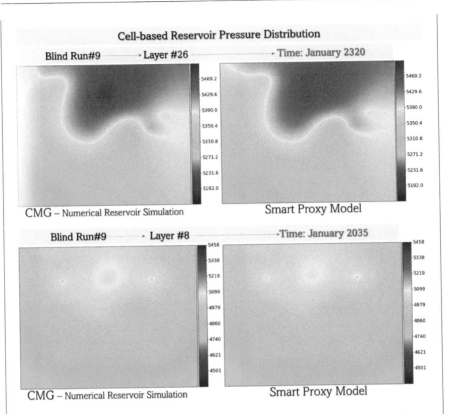

Figure 4.59 Comparing the results of NRS with the results generated by Smart Proxy Model for a specific time-step of the reservoir pressure distribution in saline aquifer. Blind validation (geological realization) run number 9; top figure: Layer 26 in January 2320, and run number 9; bottom figure: Layer 8 in January 2035.

number 13 for layer 26 in time-step December 2022, and layer 40 in time-step April 2020. Figure 4.61 shows the reservoir pressure distribution of blind validation geological realization and its NRS run number 17 for layer 41 in time-step September 2021, and layer 20 in time-step January 2049.

In this section, the reservoir pressure distribution in space and time for every cell in blind validation geological realization and its NRS is being compared; several comparisons in 2-D and 3-D plots are shown. Figures 4.62–4.65 show reservoir pressure distribution in space and time for multiple time-steps of a specific layer of a blind validation geological realization and its NRS run. Figure 4.62 shows the comparison between the commercial NRS of the blind geological

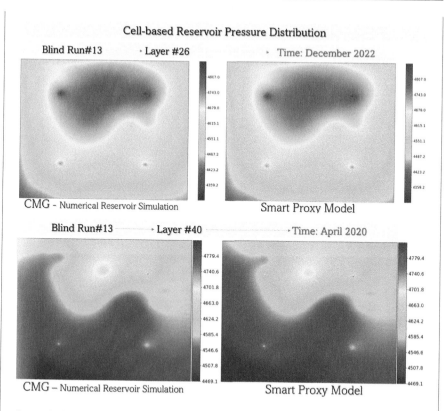

Figure 4.60 Comparing the results of NRS with the results generated by Smart Proxy Model for a specific time-step of the reservoir pressure distribution in saline aquifer. Blind validation (geological realization) run number 13; top figure: Layer 26 in December 2022, and run number 13; bottom figure: Layer 40 in April 2020.

realization number 9 for layer number 5. Since the dynamic Smart Proxy Model was deployed and generated the entire time-steps and all layers, this figure (Figure 4.62) shows four time-steps (February 2020, August 2020, June 2021, and June 2022) for layer 5.

Figure 4.63 shows the reservoir pressure distribution comparison between the commercial NRS of the blind geological realization number 9 for layer 5 in another four time-steps (January 2025, January 2030, January 2035, and January 2040) for layer 5. Figure 4.64 shows the reservoir pressure distribution comparison between the commercial NRS of the blind geological realization number 9 for layer 5 in another four time-steps (January 2045, January 2050, January 2055, and January 2060) for layer 5, and finally, Figure 4.65 shows the reservoir

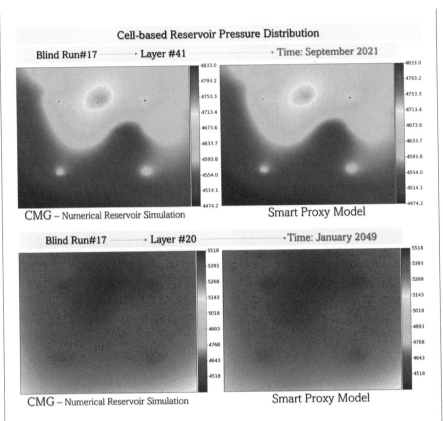

Figure 4.61 Comparing the results of NRS with the results generated by Smart Proxy Model for a specific time-step of the reservoir pressure distribution in saline aquifer. Blind validation (geological realization) run number 41; top figure: Layer 26 in September 2021, and run number 17; bottom figure: Layer 20 in January 2049.

pressure distribution comparison between the commercial NRS of the blind geological realization number 9 for layer 5 in another four time-steps (January 2070, January 2100, January 2120, and January 2320) for layer 5.

Finally, 3-D comparison of reservoir pressure distribution of commercial NRS of the blind geological realization and dynamic Smart Proxy Model are shown in Figures 4.66–4.69. Each of these figures shows the 3-D reservoir pressure distribution both from the top and the bottom for all the simulation layers in a given time-step. Figure 4.66 shows the 3-D view of the reservoir pressure distribution comparison of NRS and dynamic Smart Proxy Model for the entire field for blind validation run number 5 in November 2020. Figure 4.67 shows

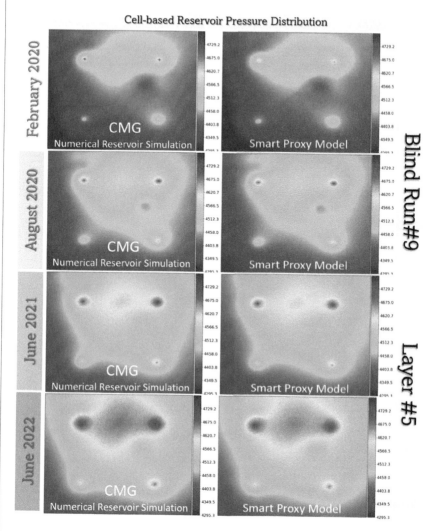

Figure 4.62 Comparing reservoir pressure distribution of NRS with dynamic Smart Proxy Model for a specific layer. The dynamic Smart Proxy model represents the entire field at all time-steps. This figure shows layer 5 of blind validation number 9 in February and August 2020, June 2021, and June 2022.

the 3-D view of the reservoir pressure distribution comparison of NRS and dynamic Smart Proxy Model for the entire field for blind validation run number 5 in January 2050.

Figure 4.68 shows the 3-D view of the reservoir pressure distribution comparison of NRS and dynamic Smart Proxy Model for the entire field for blind validation run number 13 in February 2020, and Figure 4.69 shows the 3-D

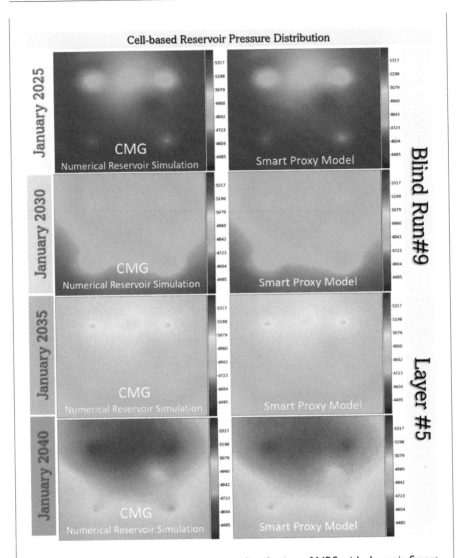

Figure 4.63 Comparing reservoir pressure distribution of NRS with dynamic Smart Proxy Model for a specific layer. The dynamic Smart Proxy model represents the entire field at all time-steps. This figure shows layer 5 of blind validation number 9 in January 2025, 2030, 2035, and 2040.

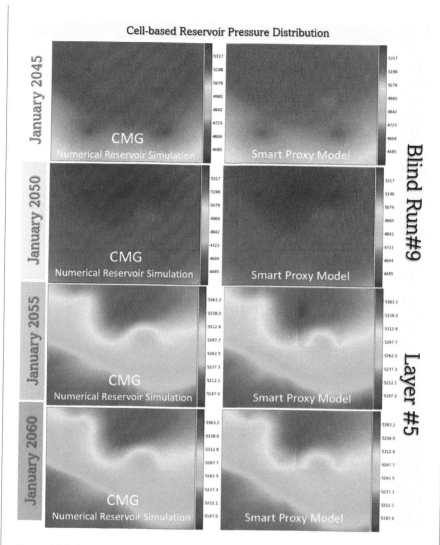

Figure 4.64 Comparing reservoir pressure distribution of NRS with dynamic Smart Proxy Model for a specific layer. The dynamic Smart Proxy model represents the entire field at all time-steps. This figure shows layer 5 of blind validation number 9 in January 2045, 2050, 2055, and 2060.

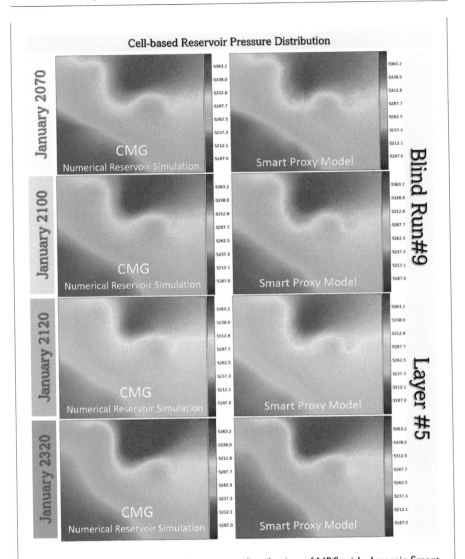

Figure 4.65 Comparing reservoir pressure distribution of NRS with dynamic Smart Proxy Model for a specific layer. The dynamic Smart Proxy Model represents the entire field at all time-steps. This figure shows layer 5 of blind validation number 9 in January 2070, 2100, 2120, and 2320.

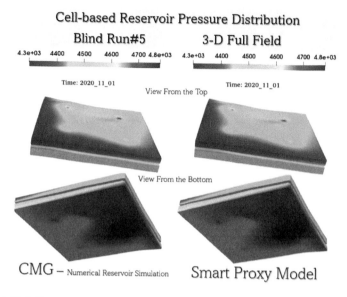

Figure 4.66 3-D view of the reservoir pressure distribution comparison of NRS and dynamic Smart Proxy Model for the entire field at all time-steps. This figure shows the 3-D view from the top and the bottom for blind validation run number 5 in November 2020.

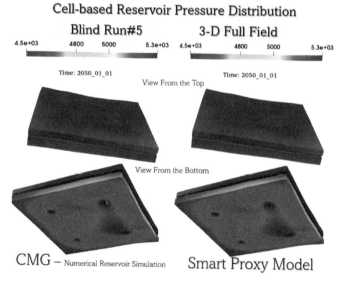

Figure 4.67 3-D view of the reservoir pressure distribution comparison of NRS and dynamic Smart Proxy Model for the entire field at all time-steps. This figure shows the 3-D view from the top and the bottom for blind validation run number 5 in January 2050.

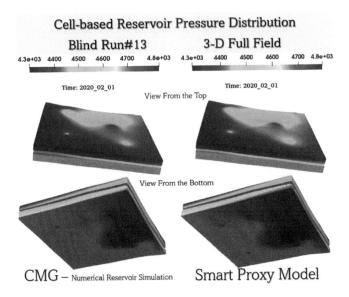

Figure 4.68   3-D view of the reservoir pressure distribution comparison of NRS and dynamic Smart Proxy Model for the entire field at all time-steps. This figure shows the 3-D view from the top and the bottom for blind validation run number 13 in February 2020.

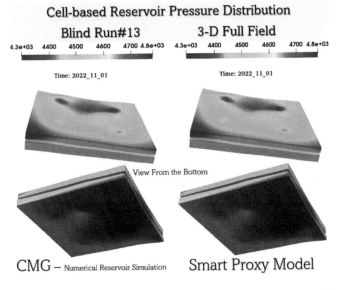

Figure 4.69   3-D view of the reservoir pressure distribution comparison of NRS and dynamic Smart Proxy Model for the entire field at all time-steps. This figure shows the 3-D view from the top and the bottom for blind validation run number 13 in November 2022.

view of the reservoir pressure distribution comparison of NRS and dynamic Smart Proxy Model for the entire field for blind validation run number 13 in November 2022.

CO_2 SATURATION RESULTS OF THE CO_2 STORAGE SMART PROXY MODEL

Smart Proxy Modeling results of CO_2 plume are presented in the next eight figures (Figures 4.70–4.77). Figure 4.70 shows the CO_2 saturation distribution of blind validation geological realization and its NRS run number 5 for layer 6 in time-step January 2050, and for layer 9 in time-step January 2050. Figure 4.71 shows the CO_2 saturation distribution of blind validation geological realization and its NRS run number 5 for layer 7 in time-step January 2070, and for layer 14 in time-step January 2070.

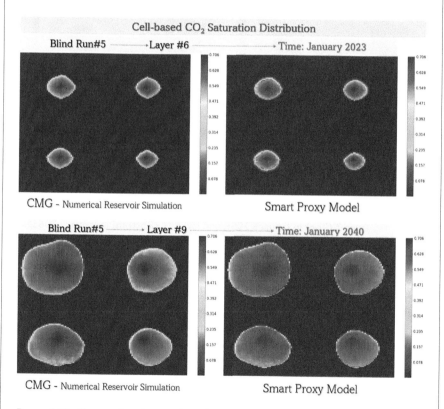

Figure 4.70 Comparing the results of NRS with the results generated by Smart Proxy Model for a specific time-step of the CO_2 plume distribution in saline aquifer. Blind validation (geological realization) run number 5 for layer 6 (Top) in January 2023, and layer 9 (Bottom) in January 2040.

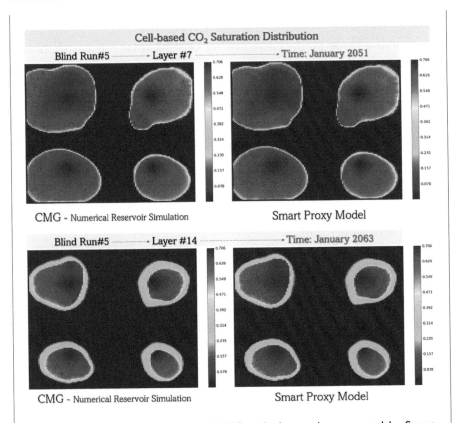

Figure 4.71 Comparing the results of NRS with the results generated by Smart Proxy Model for a specific time-step of the CO_2 plume distribution in saline aquifer. Blind validation (geological realization) run number 5 for layer 7 (Top) in January 2051, and layer 14 (Bottom) in January 2063.

Figures 4.72–4.75 show CO_2 saturation distribution in space and time for multiple time-steps of a specific layer of a blind validation geological realization and its NRS run number 5. Figure 4.72 shows the comparison between the commercial NRS of the blind geological realization number 5 for layer 6. Since the dynamic Smart Proxy Model was deployed and generated the entire time-steps and all layers, this figure (Figure 4.72) shows four time-steps (January 2023, January 2033, January 2040, and January 2050) for layer 6. Figure 4.73 shows the comparison between the commercial NRS of the blind geological realization number 5 for layer 7 in another four time-steps (January 2051, January 2060, January 2066, and January 2070) for layer 7.

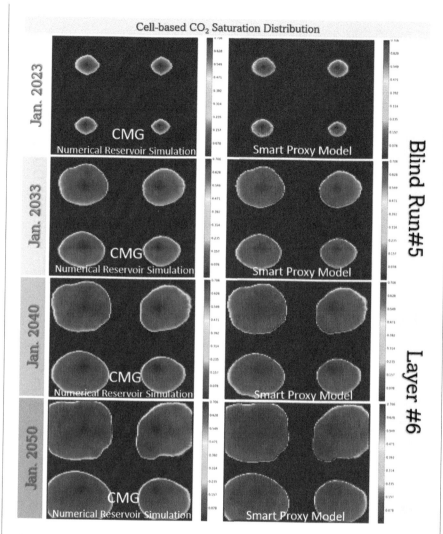

Figure 4.72 Comparing CO_2 plume distribution of NRS with dynamic Smart Proxy Model for a specific layer. The dynamic Smart Proxy Model represents the entire field at all time-steps. This figure shows layer 6 of blind validation number 5 in January 2023, 2033, 2040, and 2050.

Figure 4.74 shows the CO_2 saturation distribution comparison between the commercial NRS of the blind geological realization number 5 for layer 9 in another four time-steps (January 2026, January 2039, January 2045, and January 2050) for layer 9, and finally Figure 4.75 shows the CO_2 saturation distribution comparison between the commercial NRS of the blind geological realization

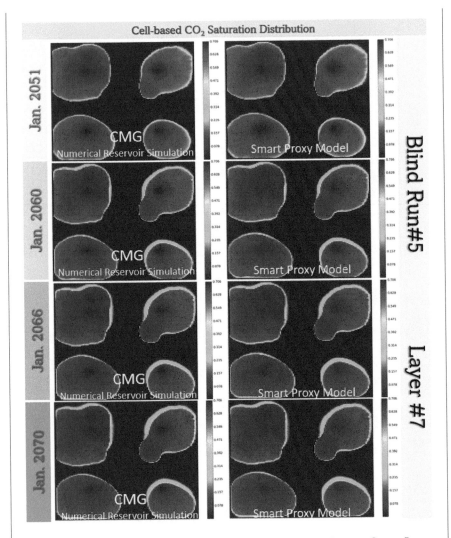

Figure 4.73 Comparing CO_2 plume distribution of NRS with dynamic Smart Proxy Model for a specific layer. The dynamic Smart Proxy Model represents the entire field at all time-steps. This figure shows layer 7 of blind validation number 5 in January 2051, 2060, 2066, and 2070.

number 5 for layer 14 in another four time-steps (January 2052, January 2055, January 2063, and January 2070) for layer 9. Finally, 3-D comparison of CO_2 saturation distribution of commercial NRS of the blind geological realization and dynamic Smart Proxy Model are shown in Figures 4.76 and 4.77.

These two figures show the 3-Dview of the CO_2 plume distribution both from the top and the bottom for all the simulation layers in a given time-step.

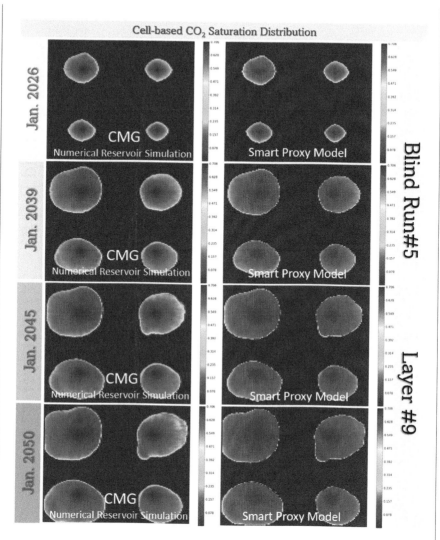

Figure 4.74 Comparing CO_2 plume distribution of NRS with dynamic Smart Proxy Model for a specific layer. The dynamic Smart Proxy Model represents the entire field at all time-steps. This figure shows layer 9 of blind validation number 5 in January 2026, 2039, 2045, and 2050.

Figure 4.76 shows the 3-D view of the reservoir pressure distribution comparison of NRS and dynamic Smart Proxy Model for the entire field for blind validation run number 5 in November 2020. Figure 4.67 shows the 3-D view of the reservoir pressure distribution comparison of NRS and dynamic Smart Proxy Model for the entire field for blind validation run number 5 in January 2050.

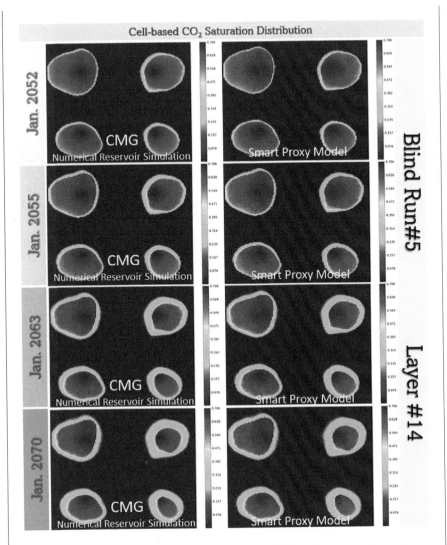

Figure 4.75 Comparing CO_2 plume distribution of NRS with dynamic Smart Proxy Model for a specific layer. The dynamic Smart Proxy Model represents the entire field at all time-steps. This figure shows layer 14 of blind validation number 5 in January 2052, 2055, 2063, and 2070.

Figure 4.76 shows the 3-D view of the CO_2 plume distribution comparison of NRS and dynamic Smart Proxy Model for the entire field for blind validation run number 5 in January 2023, and Figure 4.77 shows the 3-D view of the CO_2 plume distribution comparison of NRS and dynamic Smart Proxy Model for the entire field for blind validation run number 5 in January 2033.

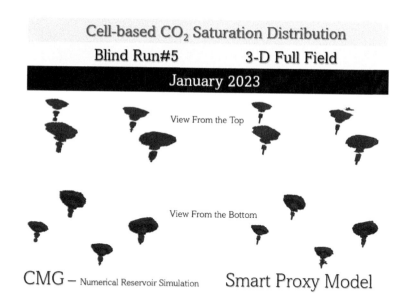

Figure 4.76 3-D view of the CO_2 plume distribution comparison of NRS and dynamic Smart Proxy Model for the entire field at all time-steps. This figure shows the 3-D view from the top and the bottom for blind validation run number 5 in January 2023.

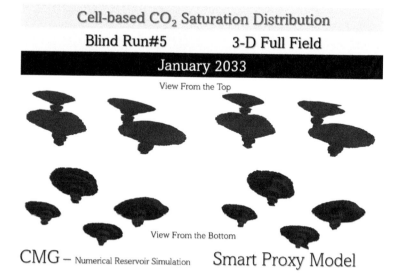

Figure 4.77 3-D view of the CO_2 plume distribution comparison of NRS and dynamic Smart Proxy Model for the entire field at all time-steps. This figure shows the 3-D view from the top and the bottom for blind validation run number 5 in January 2033.

Chapter 5

Smart Proxy Modeling for computational fluid dynamics (CFD)

Smart Proxy Modeling can be applied to any type of numerical simulation. In this chapter, application of Smart Proxy Modeling to computational fluid dynamics (CFD) is covered. All details about the definitions and development of Smart Proxy Models that have already been covered in the previous chapters of this book are applicable to the application of Smart Proxy Model to CFD. In this chapter, the Smart Proxy Model is applied to a CFD numerical simulation that is applied to Tri-State boiler. Instead of the results being done as a function of time and space, the results were generated and being covered on a particular time that would represent the process when it comes to a common and continued behavior.

To fulfill the prescriptive analytics and combustion optimization objectives of the industrial scale Tri-State boiler model, simulation models to be used in developing the Smart Proxy Models must be developed using a realistic set of boundary conditions that are representative of the actual Tri-State boiler operating conditions. Considering the complex relationship among the several controllable variables that impact the characteristics of the resulting reaction flow and heat transfer process, a set of "Tri-State Boiler" CFD simulation runs was generated with a simplified set of boundary conditions.

The objective is to develop a Smart Proxy Model that can replicate the Tri-State boiler CFD simulation results of combustion transport variables of interest. These transport variables include the gas pressure, temperature, nitrogen concentration, carbon dioxide concentration, and oxygen concentration at different flow rates of coal and air (primary and secondary) at the injection ports.

Case study: Smart proxy modeling of tri-state boiler computational fluid dynamics (CFD)

Co-authors of this specific section of this chapter are Yvon Martinez, Daniel Keller, Ayodeji Aboaba from West Virginia University Laboratory for Engineering Application of Data Science (WVU-LEADS), and

DOI: 10.1201/9781003242581-5

Dr. Mehrdad Shahnam, Chris Guenther from National Energy Technology Laboratory (NETL) of US Department of Energy, and Yong Liu from Leidos.

With the growing need for cleaner and more efficient energy, power plants need a reliable tool for optimizing power output. Industrial, commercial, and institutional (ICI) boilers, for example, require tools centered around the study of computational fluid dynamics to understand the thermal-flow patterns and species distribution of coal-fired combustion in a high-pressure facility. Because the necessary calculations can be computationally expensive and time-consuming, Smart Proxy technology was used to assist in predictive analysis of ICI boilers.

Smart Proxy technology leverages Artificial Intelligence and Machine Learning to discover hidden and predictable patterns within phenomena in a manner that is significantly faster and computationally cheaper than using traditional numerical simulators. Smart Proxy Modeling organizes data in order to teach a Machine Learning algorithm to predict specified outputs for given inputs. In this process, fuel-air mixture and boiler geometry are given as inputs, and the pressure, temperature, and levels of oxygen, carbon dioxide, and nitrogen of the exhaust are given as target outputs. By successfully teaching the Machine Learning algorithm the physics of dealing with ICI boilers, the typical long time-to-solution characteristics of CFD simulations are significantly reduced while preserving traditional CFD solver accuracy for the model under study.

INTRODUCTION OF TRI-STATE BOILER SMART PROXY

Electricity generation by coal is one of the most important activities in fossil-fuel-based economies across the globe. Despite its significance, the use of coal for electricity generation poses adverse impacts on humans and the environment, especially excessive emissions of greenhouse gases (GHGs) to the atmosphere. In recent years, the increasing demands for power generation and stringer environmental requirements have motivated implementation in coal-fired power generations optimization analysis [Gu et al., 2018]. An improved efficiency strategy can be realized through either adjusting the power plant operating conditions and or modifying the plant's configuration to fully utilize energy resources within the system [Sanpasertparnich and Aroonwilas, 2009]. A typical coal-fired power plant consists of boiler, turbine, and electric generator devices. The operating condition of boiler greatly impacts the total efficiency of the power plant unit. Coal-fired power plants utilize high-temperature and high-pressure steam generated by the combustion of pulverized coal in the boiler to power steam turbine rotors in order to generate electricity. CFD simulation technologies have been employed to understand the thermal flow and coal-air combustion phenomena in boilers to resolve operation problems and search for optimal solutions [Saripalli et al., 2005].

The purpose of developing CFD simulations of multi-phase flows is to analyze the intricate details on a particle physics scale as a function of time. Due to the nature of particle behaviors and the level of precision needed for convergence of a solution, the grids or individual points of calculations should be spaced at least ten times the particle diameter apart [Fullmer and Hrenva, 2016]. This level of resolution requires excessive computational resources [Shahnam et al., 2016]. Meaningful and practical solutions have been found in the National Energy Technology Laboratory (NETL) work after a large number of simulation runs, but this demand for computational energy is too high for many projects and design analysts [Ansari, 2016; Ansari et al., 2017; Shahnam et al., 2016].

The goal of this case study is to demonstrate how to develop an AI-based predictive model capable of replicating the thermal-flow pattern and species distribution results of CFD simulation of gas-solid reaction flows exhibited in coal combustion boilers. Achieving this goal will significantly reduce the typical long time-to-solution characteristics of CFD simulations. While preserving the results of the traditional CFD accuracy, it enables efficient design of uncertainty quantification and optimization of the model.

In this case study, details of the research performed in developing predictive models targeted toward the simulation of combustion reactions specifically for the Tri-State coal-fired boiler are presented. The beginning sections of this case study introduce the challenge and objective needed to be reached. Here, the overarching purpose of the project as well as the field is discussed, and justification for this endeavor is established. Background information concerning the project and research tools, or methods, is given as a general precursor to the methodology.

The background covers the coal-fired boiler of interest, including a brief explanation of the mechanisms as well as how the boiler is being simulated. The background also touches Machine Learning techniques used to solve the problem given. Previous work is given to establish the validity of the technology as well as introduced diverse use cases. After describing the CFD simulation modeling mechanics, this case study provides a detailed description of the workflow and thought process for decisions made for this Smart Proxy development. The methodology covers all the relevant features, both native and engineered, as well as the steps taken to enhance the model. The model was developed in steps and the enhancement steps are discussed to provide narration on the purpose of their implementation.

ESCALANTE TRI-STATE COAL POWER PLANT

Escalante is a coal-fired generating station located in Prewitt, New Mexico, 27 miles northwest of Grant, New Mexico in the heart of uranium mining area about 105 miles west of Albuquerque, New Mexico. Escalante was designed by Burns & McDonnell, based in Kansas City, Missouri. The station went into commercial operation in 1985. Coal is the fuel used to heat water, which creates steam and moves the blades of a turbine to generate electricity. Coal is shipped by rail from the Lee Ranch Mine, which is located 37 miles from the Escalante station. The plant uses approximately 800,000 tons of coal per year.

The plant is equipped with a five-level tangentially fired coal boiler. The net capacity is 245 megawatts. The boiler burns New Mexico sub-bituminous coal at about 9000 Btu/lb. Five pulverizers in the unit deliver pulverized coal at five levels, at four corners of the boiler. The pulverizers grind coal to a fine powder to ensure better mixing and combustion in the boiler. A total of 99.9% of the fly ash (a by-product of burning coal) is captured by the plant's bag-house system. A wet limestone scrubber system removes 95% of the sulfur dioxide emissions. The plant is located 6,900 feet above main sea level.

BOILER FUNCTION

INJECTORS

Five "Coal" ports were utilized to inject both pre-heated air and pulverized coal. Three "CFS" (Concentric Fire System) ports with the capability to adjust the angles of the nozzles were placed between each coal port to direct air toward the furnace walls to reduce fouling and produce an oxidizing environment along the water walls. Additional air was introduced through three "CCOFA" (Close-Coupled Over Fire Air) ports set above the top coal burners for NOx reduction. Three of the injection ports were closed as these are primarily used for the injection of oil and gas fuels that are not relevant to this study.

WATER CIRCULATION

Before water can be used in the boiler process, it must be treated and cleaned of minerals such as iron or calcium as well as any other contaminants. The presence of contaminants can damage piping via abrasion or cause buildup due to water evaporation. The treated water is typically preheated when it enters the system using feeder tanks [Bases, 2017]. The water enters in what is known as the economizer. As the water circulates through economizer, it heats up [Bases, 2017]. Upon exiting the economizer, water flows through a series of pipes toward steam drum. The steam drum is a cylinder that collects and distributes steam as it circulates throughout the boiler [Bases, 2017].

The water naturally flows down a series of pipes toward the furnace walls and convection pass where its intense heat raises its temperature [Bases, 2017]. The water transitions to steam and begins to rise through a set of pipes toward steam drum at the top of the boiler. Before the steam continues to the next circulation phase, it must be dried [Bases, 2017]. Given that the steam and water are circulating together in the steam drum, the steam at this stage is very wet and dense. The steam moves through cyclone separators that use centrifugal force to allow for lighter dry steam to continue forward, while the water and wet steam remain [Bases, 2017]. The water and wet steam continue circulating the boiler until they become dry enough to pass through the separator. The dry steam then advances toward superheater.

SUPERHEATERS

The superheater is a section of looped tubing where the dry steam flows through and is further heated by the exhaust flue gases produced from coal and air combustion [Bases, 2017]. The superheater section serves as the primary tool for temperature control [Bases, 2017]. As the steam circulates through the tubing, attemperators spray wet steam on the tubing to control the temperature of the steam. Subsequent superheater regions are sometimes referred to as reheaters. Reheaters function like a primary superheater but provide a further degree of control over the temperature of the steam. After the steam flows through the superheaters and reheaters, it then flows into the economizer.

ECONOMIZERS

Economizer is the heat exchanger that raises the temperature of the water leaving the highest-pressure feed-water heater to the saturation temperature corresponding to the boiler pressure, which is done by the gases leaving the last superheater or reheater that still has enough heat to transfer before it leaves to the stack, and that is why it is called economizer. Economizers come plain or with extended surfaces to enhance heat transfer. They are generally placed between the last super heater reheater and the air pre-heater. Economizers function better with feed-water heater, as without them cold water enters the economizer, which results in condensation and corrosion at outer surfaces.

ANSYS FLUENT

The CFD model is based on the mass, momentum, and energy balance equations with some other constitutive equations such as the equation of state to calculate the gas phase density.

$$\rho_g = \frac{PM_{wg}}{RT_g} \tag{5.1}$$

Continuity equation:

$$\frac{\partial}{\partial t}\left(\rho_g\right) + \nabla \cdot \left(\rho_g \vec{u_g}\right) = \sum_{i=1}^{N} R_{gi} \tag{5.2}$$

Momentum equation:

$$\frac{\partial}{\partial t}\left(\rho_g \vec{u_g}\right) + \nabla \cdot \left(\rho_g \vec{u_g}\vec{u_g}\right) = -\nabla P_g + \nabla \cdot \overline{\overline{\tau}}_g + \rho_g \vec{g} \tag{5.3}$$

$$\overline{\overline{\tau}}_g = \mu_e\left[\left(\nabla \vec{u_g} + \nabla \vec{u_g}^T\right) - \frac{2}{3}\nabla \cdot \vec{u_g}I\right] \tag{5.4}$$

As the gas phase is composed of several components such as the O_2, N_2, CO_2, etc., the species transport equation can be given as

$$\frac{\partial}{\partial t}\left(\rho_g Y_i\right) + \nabla \cdot \left(\rho_g \vec{u_g}Y_i\right) = -\nabla \cdot \vec{J_i} + R_{gi} \tag{5.5}$$

$$\vec{J_i} = -\left(\rho_g D_{im} + \frac{\mu_t}{Sc_t}\right)\nabla Y_i \tag{5.6}$$

where:

ρ_g = gas density (kg/m³)
P = operating gas pressure (outlet pressure)
M_w = average molecular weight of gas
R = universal gas constant (8.314 J/mol/K)
T = gas phase temperature
$\vec{u_g}$ = gas phase velocity in x, y, and z directions, respectively
$\overline{\overline{\tau}}_g$ = stress tensor
Y_i = fraction of species I in the gas phase
R_{gi} = net rate of production of species i by chemical reaction
$\vec{g}$ = gravity
$\vec{J_i}$ = diffusion flux of species due to the gradients of concentration
D_{im} = mass diffusion coefficient for species j in the mixture.
μ_e = effective viscosity ($\mu_e = \mu_t + \mu$)
μ_t = turbulent viscosity ($\mu_t = \rho C_\mu \frac{k^2}{\varepsilon}$)

Subscript "g" means the gas phase, subscript "i" means the species i.

Incompressible ideal gas law is used to calculate the gas density as the temperature changes a lot but the pressure changes little. The operating pressure is the pressure at the coal boiler outlet. For turbulent flows, the molecular viscosity is much smaller than the turbulent viscosity.

TURBULENCE MODEL

Realizable k-ε with standard wall functions as realizable k-ε model are more suitable for flow with swirling [ANSYS, Inc., 2015; Chen et al., 2017; Ge et al., 2017; Gubba et al., 2012; Modliński et al., 2015; Sun et al., 2016;

Zhou et al., 2014]. The realizable k-ε model differs from the standard k-ε model in two important ways: The realizable model contains an alternative formulation for the turbulent viscosity. A modified transport equation for the dissipation rate, has been derived from an exact equation for the transport of the mean-square vorticity fluctuation. The term "realizable" means that the model satisfies certain mathematical constraints on the Reynolds stresses, consistent with the physics of turbulent flows. Neither the standard k-ε model nor the RNG k-ε model is realizable.

The difference between the realizable k-ε model and the standard and RNG k-ε models is that C_μ is no longer constant but a function of the mean strain and rotation rates, the angular velocity of the system rotation, and the turbulence fields.

k is the turbulence kinetic energy.

$$\frac{\partial}{\partial t}\left(\rho_g k\right) + \nabla \cdot \left(\rho_g k \vec{u_g}\right) = \nabla \cdot \left[\left(\mu + \frac{\mu_t}{\sigma_k}\right)\nabla k\right] + G_k + G_b - \rho\varepsilon - Y_m + S_k \qquad (5.7)$$

ε is the dissipation rate of turbulence kinetic energy.

$$\frac{\partial}{\partial t}\left(\rho_g \varepsilon\right) + \nabla \cdot \left(\rho_g \varepsilon \vec{u_g}\right) =$$

$$\nabla \cdot \left[\left(\mu + \frac{\mu_t}{\sigma_\varepsilon}\right)\nabla \varepsilon\right] + \rho C_1 S \varepsilon - \rho C_2 \frac{\varepsilon^2}{k + \sqrt{\upsilon\varepsilon}} + C_{1\varepsilon}\frac{\varepsilon}{k}C_{3\varepsilon}G_b + S_\varepsilon \qquad (5.8)$$

where:

G_k = generation of turbulence kinetic energy due to the mean velocity gradients

G_b = generation of turbulence kinetic energy due to buoyancy

Y_m = contribution of the fluctuating dilatation in compressible turbulence to the overall dissipation rate

C_2 (1.9) and $C_{1\varepsilon}$ (1.44) = constants

σ_k (1.0) and σ_ε (1.2) = turbulent Prandtl numbers for k and ε, respectively

S_k and S_ε = user-defined source terms

CFD REACTION EDDY-DISSIPATION MODEL

Most fuels are fast burning, and the overall rate of reaction is controlled by turbulent mixing. The net rate of production of species due to reaction $R_{i,r}$ is given by the smaller (that is, limiting) value of the two expressions below:

$$R_{i,r} = \upsilon'_{i,r} M_{w,i} 4.0\rho \frac{\varepsilon}{k} \min_R \left(\frac{Y_R}{\upsilon'_{R,r} M_{w,R}}\right) \qquad (5.9)$$

$$R_{i,r} = v'_{i,r} M_{w,i} 2.0\rho \frac{\varepsilon}{k} \frac{\sum_P Y_P}{\sum_j^N v''_{j,r} M_{w,j}}$$ (5.10)

where:

Y_P = mass fraction of any product species, P

Y_R = mass fraction of a reactant, R

$v'_{i,r}$ = stoichiometric coefficient for reactant i in reaction r

$v''_{i,r}$ = stoichiometric coefficient for product i in reaction r

CFD HEAT TRANSFER MODEL

$$\frac{\partial}{\partial t}(\rho_g H) + \nabla \cdot (\rho_g H \vec{u_g}) = \nabla \cdot \left(\frac{k_t}{C_p} \nabla H \right) + S_h$$ (5.11)

$$H_j = \int_{T_{ref}}^{T} C_{p,j} dT + H_j^0 (T_{ref,j})$$ (5.12)

H is the total enthalpy defined as $H = \Sigma_j Y_j H_j$, where Y_j is the mass fraction of species j and H_j is the enthalpy of species j. The heat capacity $C_{p,j}$ is defined as a function of temperature for each species. When the radiation model is being used, the source term S_h includes radiation source terms. Both conduction and convection require matter to transfer heat. Radiation is a method of heat transfer that does not rely upon any contact between the heat source and the heated object. Thermal radiation (often called infrared radiation) is a type of electromagnetic radiation (or light). Radiation is a form of energy transport consisting of electromagnetic waves traveling at the speed of light. No mass is exchanged, and no medium is required for radiation.

RADIATION MODEL

Discrete ordinates (DO) model is used as DO model needs more computational resource than other radiation model, but DO model is more complete [ANSYS, Inc., 2015; Chen et al., 2017; Choi, 2009; Edge et al., 2011; Ge et al., 2017; Gubba et al., 2012; Modliński et al., 2015; Zhou et al., 2014].

DO is recommended by "ANSYS Fluent". Emissivity of gas can be calculated from weighted-sum-of-grey-gases model (WSGGM), which has been widely used in CFD and maintains a good balance between calculating efficiency and accuracy [Smith et al., 1982; Yin, 2013]. WSGGM assumed that the emissivity of flue gas was decided by local temperature and partial pressure of gas species.

SUMMARY OF THE NUMERICAL SIMULATION (CFD) MODEL

A numerical simulation involves the creation of a mesh that represents a geometric domain in a three-dimensional (3-D) space. The time of convergence to a particular solution can take hours, if not days, depending on the complexity and the type of mesh.

MESHING

A structured mesh of dimensions 13.56 m wide, 10.23 m deep, and 43.62 m high was generated from the aggregation of 4,748,328 tetrahedron cells to model the physics that a coal power plant boiler typically undergoes (Figure 5.1). Every cell in the system was constructed from a group of nodes and a centroid, both with known x, y, and z coordinates, which determine its size and volume. For example, Figure 5.2 shows a single tetrahedron cell structure formed by a series of four interconnected nodes and a geometric center point known as cell centroid.

REGIONS

Due to geometry complexities as well as the addition of internal heat exchanger plates, the boiler was divided into seven regions (Figure 5.3). Some regions had a more refined mesh to account for sudden geometry changes or locations in which more complex reactions, such as the injection surfaces, took place.

BOUNDARY CONDITIONS

Once the mesh was generated, the boundary conditions of the CFD boiler were defined at the inlet, outlet, walls, and internal walls, as shown in Figure 5.4 [García et al., 2012]. The inlet of the boiler comprised 15 injection ports at

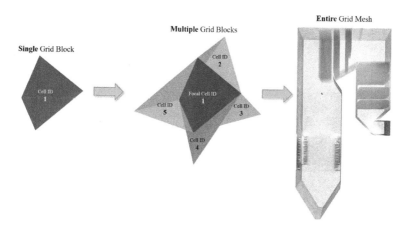

Figure 5.1 Construction of grid (mesh) for numerical simulation.

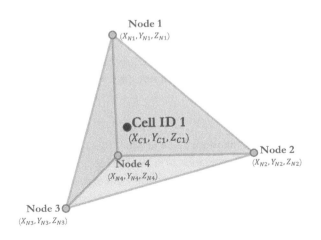

Figure 5.2 Cell geometry.

each of the four corners of the wind box. Five "COAL" ports were utilized to inject both pre-heated air and pulverized coal. Three "CFS" ports were placed between the coal ports without any angle modifications. The "CCOFA" ports were set above the top coal burners. A single auxiliary port of air (AIR AA) was positioned at the bottom. Three of the injection ports were closed as these

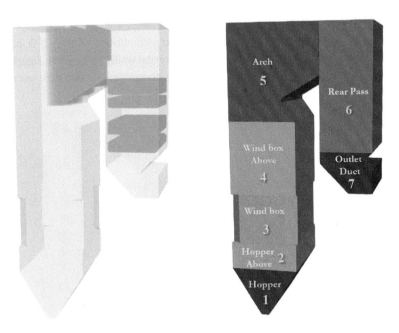

Figure 5.3 CFD boiler numerical simulation regions.

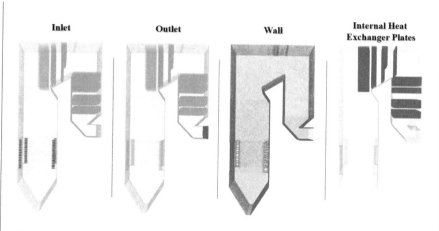

Figure 5.4 Cells located at inlet (furthest left), outlet (middle left), wall (middle right), and internal heat exchanger plates (furthest right).

were primarily used for the injection of oil and gas fuels that were not relevant to this study. Figure 5.5 illustrates the location of each of the injection levels previously discussed.

The outflow conditions were set at the boiler's outlet duct. For the wall boundary conditions, a constant temperature or heat flux was defined. Eight sets of internal heat exchanger plates were accommodated throughout the arch and rear pass of the boiler.

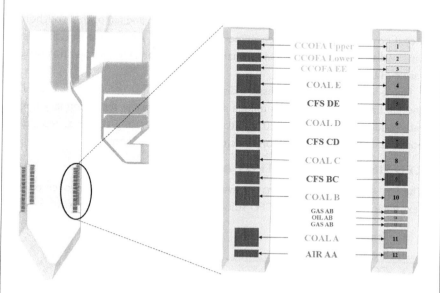

Figure 5.5 Injection ports.

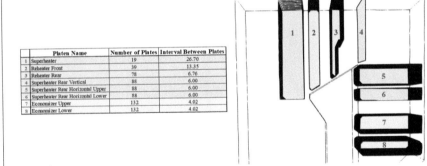

Platen Name	Number of Plates	Interval Between Plates
1 Superheater	19	26.70
2 Reheater Front	39	13.35
3 Reheater Rear	78	6.76
4 Superheater Rear Vertical	88	6.00
5 Superheater Rear Horizontal Upper	88	6.00
6 Superheater Rear Horizontal Lower	88	6.00
7 Economizer Upper	132	4.02
8 Economizer Lower	132	4.02

Figure 5.6 Platen locations.

PLATENS

The simulation model incorporated a series of zero-thickness plates throughout the boiler. Each series of plates constituted a superheater, reheater, or economizer platen. The boiler comprised four superheaters, two reheaters, and two economizer platens. The location and number of plates in each platen is summarized in Figure 5.6.

CFD SIMULATION INPUT CONDITIONS

A total of 32 steady-state CFD numerical simulations were executed through ANSYS FLUENT CFD software to determine the distribution of different output variables such as temperature, pressure, CO_2, oxygen, and nitrogen throughout a coal power plant boiler. Each numerical simulation run resulted in different output distributions when coal, primary air, and secondary air flowrates were varied (Table 5.1).

Figure 5.7 illustrates the pressure distribution for CFD simulation runs 2, 12, and 25. Each simulation has a unique set of coal, primary, and secondary flow rate conditions, ranging from 0 kg/s to 7 kg/s, depicted in green, red, and blue horizontal bars, respectively, on the chart above the boiler model.

CFD SIMULATION OUTPUT ATTRIBUTES

The final steady-state solutions for every grid block (or cell) were collected in text file format. All targeted CFD solution output data are listed below:

- Temperature
- Pressure
- CO_2
- Oxygen
- Nitrogen

Table 5.1 CFD simulation inputs

Numerical simulation run number	Coal flow rate (kg/s)	Primary air flow (kg/s)	Secondary air flow (kg/s)
1	0.81	3.28	5.83
2	1.67	3.80	4.62
3	0.71	2.70	5.41
4	1.53	3.61	2.95
5	0.97	2.80	4.30
6	1.07	3.37	4.72
7	1.30	2.56	5.69
8	1.37	2.94	4.86
9	1.24	2.75	3.19
10	1.11	2.99	5.97
11	1.14	3.75	3.88
12	1.50	2.85	6.53
13	1.40	3.70	6.11
14	1.60	3.42	5.46
15	0.74	3.47	3.33
16	0.68	3.09	4.16
17	1.17	2.37	5.13
18	0.84	2.90	3.74
19	0.94	3.18	3.05
20	1.01	3.56	5.55
21	0.88	3.66	5.00
22	0.91	2.42	6.25
23	1.27	3.51	4.44
24	1.34	3.32	3.60
25	0.65	2.32	2.91
26	1.21	3.13	6.39
27	1.04	2.61	3.47
28	1.57	2.66	4.21
29	1.63	3.04	3.37
30	1.47	3.32	5.27
31	0.78	2.51	4.58
32	1.44	2.47	4.02

SUMMARY OF THE TRI – STATE BOILER SIMULATION MODEL

In this section, a detailed description of the approach and tasks performed in developing a Smart Proxy Model for the Tri-State coal boiler model is presented. Considering that the multiphase coal boiler model is more complex

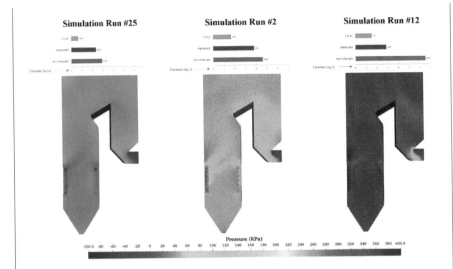

Figure 5.7 Pressure distribution at different flow rate conditions.

compared with the "B6 Combustor" that was covered in Chapter 4, the development framework becomes very important. In order to ensure the robustness and reliability of the Smart Proxy Model for use in turndown operations, prescriptive analytics and operations optimization are adopted for the "Tri-State Boiler" model; it is a two-step approach applied in developing the Smart Proxy framework for the "B6 Combustor" model.

TRI-STATE COAL COMBUSTION SIMULATION MODELS

Besides the continuity, momentum, and energy equations listed in the previous sections, there are some coal-specific models for coal combustion simulation. The reaction flow exhibited during the combustion of coal is dependent on the composition and prevailing temperature. The coal is assumed to be composed of volatile, fixed carbon, ash, and moisture.

Table 5.2 illustrates the proximate analysis of coal composition while Table 5.3 shows the elemental ultimate analysis of coal with high heating value

Table 5.2 Proximate analysis of coal (% dry weight)

Component	% Dry weight
Volatiles	33.07
Fixed carbon	53.10
Ash	13.83
Moisture	0.0

Table 5.3 Ultimate analysis of coal

Element	% Composition
Carbon	70.91
Hydrogen	4.52
Nitrogen	1.47
Sulfur	1.39
Oxygen	7.89

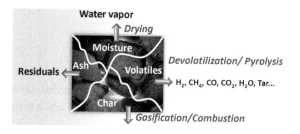

Figure 5.8 Coal combustion reaction based on composition.

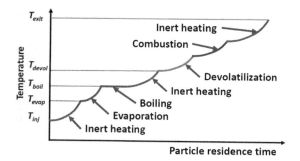

Figure 5.9 Coal combustion reactions as a function of temperature.

of 8744 Btu/lb. Figure 5.8 shows the schematic of coal composition and the corresponding combustion reaction.

Figure 5.9 shows the coal combustion reaction as a function of temperature. As coal is injected into the boiler, heated coal particles exhibit a sequence of combustion reactions (inert heating, drying, [moisture release], inert heating, devolatilization [volatile release], combustion [carbon combustion], and inert heating) depending on the temperature as shown in Figure 5.9.

The coal combustion model includes:

1. Moisture release model
2. Pyrolysis/devolatilization model mechanism; kinetics at proper heating rates/temperatures
3. Char conversion model oxidation/gasification; mass transfer; kinetics
4. Gas-phase combustion

DEVOLATILIZATION MODEL

There are many models available. For example, the two-competing-rate Kobayashi model [Hashimoto et al., 2011], i.e., one reaction controls at low temperature and the other reaction at high temperature; single rate model, etc.

The single rate model used in this simulation is

$$-\frac{dm_p}{dt} = k\left[m_p - (1 - f_{v,0})\right]m_{p,0} \tag{5.13}$$

where m_p is the particle mass (kg), $f_{v,0}$ is the mass fraction of volatile initially present in the coal, $m_{p,0}$ is the initial particle mass (kg), and k is the kinetic rate expressed as

$$k = A_1 e^{-\left(\frac{E}{RT}\right)} \tag{5.14}$$

A_1 is the pre-exponential factor and E is the activation energy.

COMBUSTION MODEL (DIFFUSION-LIMITED)

The diffusion-limited surface reaction rate model, which is the default model in ANSYS Fluent CFD commercial software, assumes that the surface reaction proceeds at a rate determined by the diffusion of the gaseous oxidant to the surface of the particle:

$$Char(s) + S_b OX(g) \rightarrow Product(g) \tag{5.15}$$

and the rate is

$$\frac{dmp}{dt} = -4\pi d_p D_{im} \frac{Y_{ox}T_\infty \rho}{Sb(T_p + T_\infty)} \tag{5.16}$$

where:
D = diffusion coefficient for oxidant in the bulk (m^2/s)
Y = local mass fraction of oxidant in the gas
ρ = gas density (kg/m^3)

GAS PHASE REACTION

Homogeneous gas fuels are fast burning, and the overall rate of homogeneous reaction is controlled by turbulent mixing. Eddy dissipation assumes that the combustion rate is much faster than turbulent dissipation and the turbulent dissipation is the limiting step [Al-Abbas et al., 2012, 2013; Bhuiyan and Naser, 2015; Ge et al., 2017; Gubba et al., 2012; Modliński et al., 2015].

The volatile composition is calculated by Fluent from the coal composition ultimate analysis. The molecular weight of volatile is assumed to be 30 kg/kmol. Coal dry density is 1400 kg/m³. Two-step reaction is used in the modeling: CO is generated from step 1 and then CO reacts with O_2 in step 2. The subscript numbers for the volatile are calculated by Fluent from the setting up of the coal calculator.

Step 1

$$C_{1.32}H_{3.56}O_{0.53}N_{0.0654}S_{0.0339} + 1.32O_2 \rightarrow 1.32CO + 1.78H_2O + 0.0327N_2 + 0.0339SO_2$$

(5.17)

Step 2

$$CO + 0.5O_2 \rightarrow CO_2$$

(5.18)

PARTICLE MOTION: DISCRETE PHASE MODEL

Unlike the Eulerian frame used for governing equations for mass, momentum, and energy in the previous section, the discrete phase model (DPM) uses the Lagrangian frame to calculate the particle track.

The DPM is useful for dilute flows when the particle volume fraction is less than 10%; however, the mass fraction may greatly exceed. The DPM traces all the particles, has detailed particle information, and is relatively faster (steady state) with reasonable particle number. The DPM can easily handle different particle diameters and there are built-in laws for moisture release, devolatilization, and char combustion.

Lagrangian particle tracking

$$\frac{dup}{dt} = F_D(u - up) + \frac{g(\rho p - \rho)}{\rho p} + F$$

(5.19)

Instantaneous momentum equation for particle

$$m_p \frac{dV_p}{dt} = \sum F$$

(5.20)

Particle trajectory equations

$$x_p = \int u_p \, dt$$

(5.21)

$$y_p = \int v_p \, dt$$

(5.22)

$$z_p = \int w_p \, dt$$

(5.23)

PROPERTIES MODEL

Besides the governing equations for mass, momentum and energy, some models are used to calculate the constitutive properties such as density, heat capacity, etc.

1. Gas phase density: Incompressible ideal gas, as the pressure does not change much in the boiler but the temperature changes a lot.

$$\rho_g = \frac{PM_{wg}}{RT_g} \tag{5.24}$$

P is the operating gas pressure (outlet pressure); M_w is the average molecular weight of gas; R is the universal gas constant (8.314 J/mol/K); and T is the gas phase temperature.

2. Heat capacity: The coal-volatiles-air is a mixture (composed of volatile, sulfur dioxide, nitrogen, water vapor, carbon dioxide, carbon monoxide, oxygen), that needs to define the property for the mixture. Mixing law is used for the mixture with polynomial or piecewise polynomial for items mentioned above.

For each species, Fluent automatically assigns the polynomial for heat capacity, and then the mixing law is used to calculate the mixture property. For example,

   - Volatile heat capacity as function of temperature: $2005 - 0.681428 * T + 0.00708589 * T2 - 4.71368 * 10 - 6 * T3 + 8.51317 * 10 - 10 * T4$
   - Sulfur dioxide: $000896 * T2 + 4.321 * 10 - 7 * T3 - 1.139 * 10 - 10 * T4$ for temperature $300 < T < 377.8587 + 1.0516 * T - 0.1000$
   - Sulfur dioxide: $681.95 + 0.2567 * T - 0.0001065 * T2 + 2.0459 * 10 - 8 * T3 - 1.454 * 10 - 12 * T4$ for temperature $1000 < T < 5000$
   - Similar piecewise polynomial formula for nitrogen and carbon monoxide
   - Similar polynomial formula for water vapor, carbon dioxide, and oxygen

3. Viscosity: Constant molecular viscosity has been used for the gaseous phase mixture as the molecular viscosity is much smaller than the turbulent viscosity and it will not affect the results a lot.

4. Thermal conductivity: Constant value has been used for the simulation.

TRI-STATE BOILER SMART PROXY DEVELOPMENT

Let's now focus on the steps taken to develop Smart Proxy Models for this particular CFD model. This section starts with the details of data collection, processing, and preparation, and covers the development, modeling, and deployment of Smart Proxy. This journey includes, but is not limited to, the

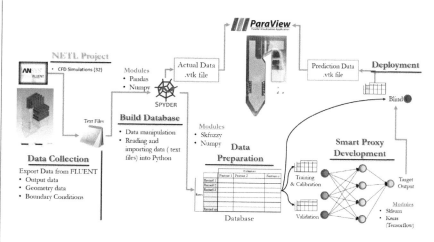

Figure 5.10 Tri-State boiler Smart Proxy Modeling workflow.

implementation, trial and error of partitioning, feature generation, sequential modeling, deep learning, and multi-output networks. Since the data was collected from a numerical simulator (Tri-State boiler CFD), it was not necessary to handle missing values and outliers.

Figure 5.10 illustrates a workflow chart that depicts an overview of the project from start to finish. The workflow started with NETL performing 32 numerical simulation runs to determine the distribution of velocities, gas species, and more. The data was provided at a cell level in text file format for all 32 simulation runs. The data was inspected, processed, and stored in a database for easy and quick access. The data was later prepared in a format to be fed to the neural network to initiate the Smart Proxy Model development. After proper training, calibration, and validation of the models developed, these were deployed on blind data to judge the performance of Smart Proxy for future use. Besides percent errors and metrics to evaluate the models, "ParaView" was a software that aided the visualization process. "ParaView" is an open-source multiple-platform application for interactive, scientific visualization.

DATA PROCESSING

The data collected from both the construction of the numerical simulation (CFD) boiler mesh and the execution process was categorized as geometry and solution output data, respectively. While the geometry data remained the same across all simulation runs performed, the solution output data varied from simulation to simulation based on coal, primary, and secondary flow rate conditions. Both sets of data described in section were processed from text file to matrix format using Python libraries such as "NumPy" and "Pandas". The result is tabular data organizes each cell ID with its corresponding coordinates, volume, nodes, and adjacent cells for all simulation runs.

FEATURE ENGINEERING

Features were generated to communicate, reinforce, and enhance the learning process of the neural network. Features with similar traits were placed in subcategories. Features and subcategories with a common theme were placed in categories.

ADJACENT AND NODE COUNT

The number of nodes and adjacent cells with respect to each focal cell was dependent on the type of mesh. The number of adjacent cells was also dependent on the wall boundary conditions set as well as the zero-thickness platens. Thus, a similar method to count encoding was used to increase the number of columns in the database. Figure 5.11 shows an example of the implementation of this method on three different focal cells. All three focal cells have four nodes. However, the number of adjacent cells varies as these cells may be next to a zero-thickness plate or a wall. In this case, a value of zero is assigned to communicate the absence of a neighboring cell.

ADJACENT VOLUMES

Adjacent volumes of the adjacent cells were added to the matrix to reinforce the information around the focal cell. Figure 5.12 shows two different tables. The first table on the left summarizes the cell volume, adjacent cells, and adjacent volumes of two different focal cells: Cell ID 1 and Cell ID 2. Each adjacent cell has a volume found in the second table on the right.

BOUNDARY CONDITIONS

A one-hot encoding approach was implemented to convert boundary conditions into integer data. For instance, a value of "1" was assigned to cells at inlet, outlet, wall, or internal walls while a value of "0" was assigned to cells that were not in any of these locations. Figure 5.13 posts two tables. The table on the left presents a list of Cell IDs with their respective boundary location. The table on the right shows

Geometry Data (13)								Feature Generation	
Grid Block (1)	Nodes in Cell (4)				Adjacent Cells (4)				Group Count (2)	
Cell ID	Node 1	Node 2	Node 3	Node 4	Adjacent 1	Adjacent 2	Adjacent 3	Adjacent 4	Node Count	Adjacent Count
1	510940	510941	510942	510943	159249	160151	569044	1565	4	4
884	511334	511336	511707	511754	813	1081	0	1070	4	3
1155	511814	511841	512029	512391	0	1441	906	433688	4	3

Figure 5.11 Feature generation: Node and adjacent count.

Geometry Data						Feature Generation			
Grid Block (1)	Cell Volume	Adjacent Cells (4)				Adjacent Volumes (4)			
Cell ID	Volume	Adjacent Cell 1	Adjacent Cell 2	Adjacent Cell 3	Adjacent Cell 4	Adjacent Volume 1	Adjacent Volume 2	Adjacent Volume 3	Adjacent Volume 4
1	0.002300	159249	160151	569044	1565	0.002543	0.002773	0.002039	0.003754
2	0.002200	210314	145076	315309	19	0.002096	0.002497	0.001969	0.001879

Cell ID	Volume
1	0.002300
2	0.002200
19	0.001879
1565	0.003754
145076	0.002497
159249	0.002543
160151	0.002773
210314	0.002096
315309	0.001969
569044	0.002039

Figure 5.12 Feature generation: Adjacent volumes.

Cell ID	Boundary Conditions
1	Internal Wall
3	Internal Wall
9	Internal Wall
15	Internal Wall
16	internal wall
5951	Wall
6293	Wall
6493	Wall
7223	Wall
7227	Wall
1209784	Outlet
1210133	Outlet
1210185	Outlet
1210231	Outlet
1210248	Outlet
2467227	Inlet
2468282	Inlet
2468506	Inlet
2468708	Inlet
2469942	Inlet

	Feature Generation			
	Boundary Conditions (4)			
Cell ID	Inlet	Outlet	Wall	Internal Wall
1	0	0	0	1
3	0	0	0	1
9	0	0	0	1
15	0	0	0	1
16	0	0	0	1
5951	0	0	1	0
6293	0	0	1	0
6493	0	0	1	0
7223	0	0	1	0
7227	0	0	1	0
1209784	0	1	0	0
1210133	0	1	0	0
1210185	0	1	0	0
1210231	0	1	0	0
1210248	0	1	0	0
2467227	1	0	0	0
2468282	1	0	0	0
2468506	1	0	0	0
2468708	1	0	0	0
2469942	1	0	0	0

Figure 5.13 Feature generation: Boundary conditions.

four columns representing the available boundary conditions. The one-hot encoding approach is expressed using binary values. For example, Cell ID 1 is located at internal walls, thus the internal wall column has a value of "1" and a value of "0" for the remaining columns. Given a scenario in which a cell was not present in any of the boundary conditions earlier mentioned, all columns were filled with zeros.

PLATENS

Additional information regarding the internal heat exchanger plates was provided by implementing an ordinal encoding approach. For instance, values "1" and "2" were assigned to cells that were on and in between the zero-thickness plates, respectively. Cells outside the internal heat exchanger plates were assigned a value of "0". Figure 5.14 provides an example of this feature encoding approach performed for one out of the eight sets of platens. The figure is focused on the superheater platen and is examined from the side and top views. Cell ID 85 is located in-between two plates, Cell ID 242 is placed next to a zero-thickness plate, and Cell ID 220 is outside the series of zero-thickness plates.

WALL BOUNDARY DISTANCES

Distances from every cell centroid to all six wall boundaries (Figure 5.15) were calculated and added to the database. This step was performed to extend the information regarding the model's geometry as well as a sense of space in the system.

INJECTION PORT DISTANCES

Distances from each focal cell to each of the injection ports were calculated. These features were generated to differentiate cells closer and further way to the combustion process. For this step, an approximate center point was

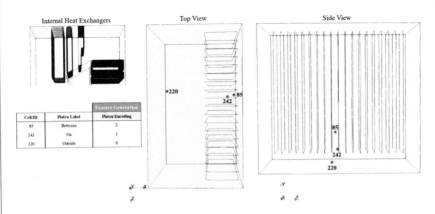

Figure 5.14 Feature generation: Platens.

	Feature Generation					
	Wall Distances (6)					
Cell ID	North	South	East	West	In	Out
1	9.6	1.1	1.9	8.4	5.4	8.2
2	0.6	10	13	8.5	12.5	1.1
3	9.6	1.1	1.9	8.4	5.1	8.4
4	9.2	0.8	1.4	9.5	6	7.5
5	8	2	3.5	9.6	5.1	8.5

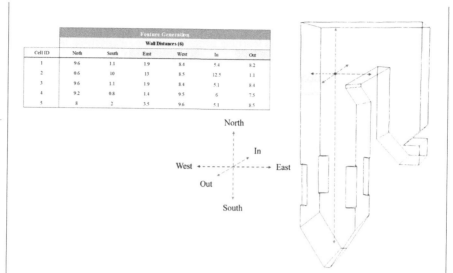

Figure 5.15 Feature generation: Wall distances.

selected from each of the 48 injection levels described in section. Figure 5.16 depicts the distance to be traveled from a focal cell located near the outlet to the first injection level at each of the four injection corners – labeled A, B, C, and D. This cell would have a larger value compared to a cell located at the furnace. The table in the figure summarizes the distances from three different focal cells to all 12 ports of coal and air at "Corner A".

	Feature Generation											
	Corner A Distances											
Cell ID	CCOFA Upper	CCOFA Lower	CCOFA EE	Coal A	Coal B	Coal C	Coal D	Coal E	CFS BC	CFS CD	CFD DE	AIR AA
1	77.02	82.02	87.02	95.52	105.72	115.92	126.12	136.32	111.42	121.62	131.82	142.52
2	85.96	90.96	95.96	104.46	114.66	124.86	135.06	145.26	120.36	130.56	140.76	151.46
3	58.83	63.83	68.83	77.33	87.53	97.73	107.93	118.13	93.23	103.43	113.63	124.33

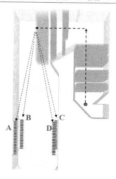

Figure 5.16 Feature generation: Injection port distances.

FUZZY CLUSTERING

To simplify the neural network's learning process and the relationship between its inputs and outputs, both fuzzy logic and Boolean logic were implemented in the feature generation process. A clustering approach was implemented to group a particular attribute of interest into total three optimum clusters. Once the best set of clusters was established, a degree of membership from each cluster was assigned to each cell. Furthermore, the highest degree of membership was identified as the main cluster through Boolean logic.

The 2-D plot in Figure 5.17 illustrates pressure values from two different CFD simulation runs. The pressure values have been grouped based on three clusters labeled C1, C2, and C3. Based on the cluster centers, a random cell colored in black, "Cell 1", does not belong entirely to the yellow cluster. Instead, the cell partly belongs to the yellow, blue, and green clusters by 62%, 27%, and 11%, respectively. Since the cell is closest to the yellow cluster, its cluster membership goes to the first cluster, C1, as summarized in the table given in Figure 5.17.

	Feature Generation					
	Degree of Membership (3)			Cluster Membership (3)		
Cell ID	DOM C1	DOM C2	DOM C3	CM C1	CM C2	CM C3
1	0.62	0.27	0.11	1	0	0
2	0.20	0.64	0.16	0	1	0
3	0.37	0.15	0.48	0	0	1

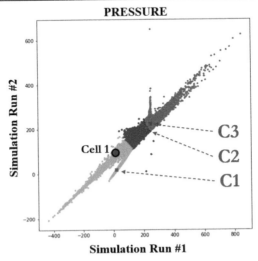

Figure 5.17 Feature generation: Fuzzy clustering.

DATA PARTITIONING (14 SIMULATION RUNS)

From the 32 simulation runs received and processed, 14 were used in the development of neural network, 6 were specifically used as blind validation, while the remaining 12 were not used in either phase, thus classified as additional blind-validation cases (Figure 5.18). The simulation runs selected for the development phase contained minimum and maximum values from nine output attributes inspected, including velocity distributions (U, V, W), gas species (CO_2, NO_2, O_2, and species volatile), pressure, and temperature. The three CFD simulation inputs and the remaining CFD simulation output variables were not taken into consideration when performing the first data partitioning.

Furthermore, the development dataset for a single attribute of interest had approximately 91 features and 66,476,592 records, from which 80% were used for training and 20% for calibration. As mentioned earlier, all CFD simulation runs were set aside for validation purposes.

NEURAL NETWORK TOPOLOGY

A feed-forward Artificial Neural Network structure (Figure 5.19) was primarily used for the development of most attributes of interest except for temperature, which did not produce satisfactory results. Figure 5.19 illustrates the respective inputs passed to the network and the output of interest to be predicted, which in this case, is pressure.

RESULTS OF THE PART ONE OF SMART PROXY MODELING

Results of the first part of Smart Proxy Modeling that took a good amount of time to be completed are presented in Figures 5.20–5.49.

Total CFD Runs			CFD Runs used for Training			Blind Validation CFD Runs		
1	12	23	1	12	23	1	12	23
2	13	24	2	13	24	2	13	24
3	14	25	3	14	25	3	14	25
4	15	26	4	15	26	4	15	26
5	16	27	5	16	27	5	16	27
6	17	28	6	17	28	6	17	28
7	18	29	7	18	29	7	18	29
8	19	30	8	19	30	8	19	30
9	20	31	9	20	31	9	20	31
10	21	32	10	21	32	10	21	32
11	22	32 Runs	11	22	14 Runs	11	22	18 Runs

Figure 5.18 Distribution of CFD simulation runs for training, calibration (14 runs), and validation (18 runs) of the Smart Proxy Model.

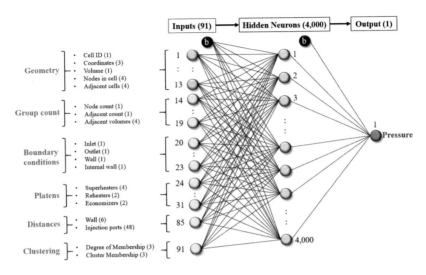

Figure 5.19 Feed-forward neural network structure.

Error Range (frac)	Simulation Run	Run Type	Number of Cells	Perc. Cells
< 0.02	6	Blind	4,642,414	97.769%
0.02 - 0.04	6	Blind	94,082	1.981%
0.04 - 0.06	6	Blind	10,217	0.215%
0.06 - 0.08	6	Blind	1,519	0.032%
0.08 - 0.1	6	Blind	73	0.002%
> 0.1	6	Blind	23	0.0%

Figure 5.20 Distribution of less than 1% CO_2 errors of the CFD blind validation run number 6.

Error Range (frac)	Simulation Run	Run Type	Number of Cells	Perc. Cells
<0.1	6	Blind	4,304,278	90.648%
0.1-0.2	6	Blind	109,387	2.304%
0.2-0.3	6	Blind	27,422	0.578%
0.3-0.4	6	Blind	15,404	0.324%
0.4-0.5	6	Blind	11,019	0.232%
0.5-0.6	6	Blind	9,487	0.2%
0.6-0.7	6	Blind	8,651	0.182%
0.7-0.8	6	Blind	8,314	0.175%
0.8-0.9	6	Blind	8,035	0.169%
> 0.9	6	Blind	241,755	5.091%

Figure 5.21 Distribution of less than 10% CO_2 errors of the CFD blind validation run number 6.

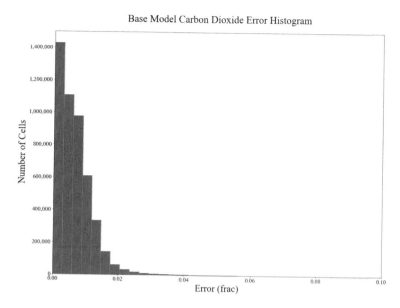

Figure 5.22 Histogram of the CO_2 errors of the CFD blind validation run number 6.

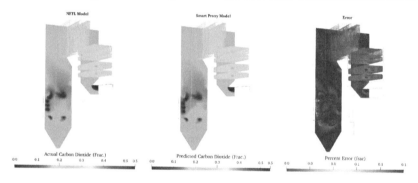

Figure 5.23 Cross section (a) of the CO_2 results – Smart Proxy of the CFD blind validation run number 6.

RESULTS OF THE FINAL VERSION OF THE CFD SMART PROXY MODELING

In this section, the final results of the CFD Smart Proxy Modeling are presented. Here the CFD Smart Proxy Modeling for temperature was enhanced.

ENHANCEMENT STEPS

The results obtained using 14 simulation runs produced satisfactory results for all outputs of interest except for temperature distribution. In this section, steps taken to enhance the temperature model are discussed.

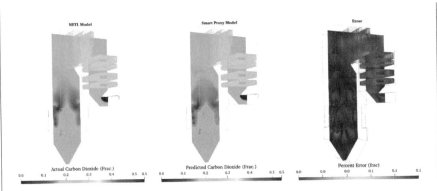

Figure 5.24 Cross section (b) of the CO$_2$ results – Smart Proxy of the CFD blind validation run number 6.

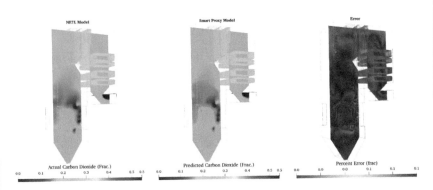

Figure 5.25 Cross section (c) of the CO$_2$ results – Smart Proxy of the CFD blind validation run number 6.

Error Range (frac)	Simulation Run	Run Type	Number of Cells	Perc. Cells
< 0.02	6	Blind	4,669,537	98.341%
0.02 - 0.04	6	Blind	69,920	1.473%
0.04 - 0.06	6	Blind	7,221	0.152%
0.06 - 0.08	6	Blind	1,460	0.031%
0.08 - 0.1	6	Blind	165	0.003%
> 0.1	6	Blind	25	0.001%

Figure 5.26 Distribution of less than 1% nitrogen errors of the CFD blind validation run number 6.

Error Range (frac)	Simulation Run	Run Type	Number of Cells	Perc. Cells
<0.1	6	Blind	4,747,355	99.98%
0.1-0.2	6	Blind	957	0.02%
0.2-0.3	6	Blind	13	0.0%
0.3-0.4	6	Blind	1	0.0%
0.4-0.5	6	Blind	1	0.0%
0.5-0.6	6	Blind	0	0.0%
0.6-0.7	6	Blind	0	0.0%
0.7-0.8	6	Blind	0	0.0%
0.8-0.9	6	Blind	0	0.0%
> 0.9	6	Blind	1	0.0%

Figure 5.27 Distribution of less than 10% nitrogen errors of the CFD blind validation run number 6.

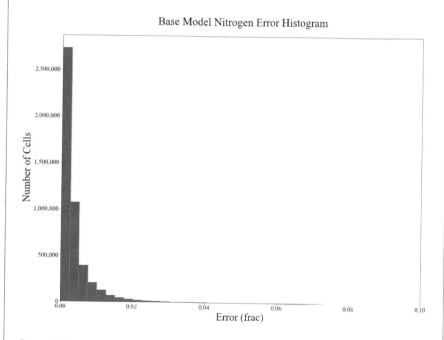

Figure 5.28 Histogram of the nitrogen errors of the CFD blind validation run number 6.

NEURAL NETWORK STRUCTURE

The objective of this enhancement was to investigate the impact of using deep learning or multiple hidden layers. In other words, we wanted to see if increasing the number of hidden layers impacts the training, calibration, and/or validation

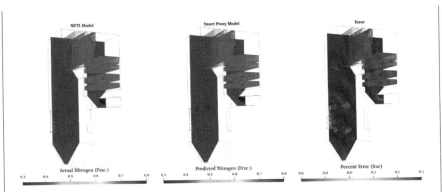

Figure 5.29 Cross section (a) of the nitrogen results – Smart Proxy of the CFD blind validation run number 6.

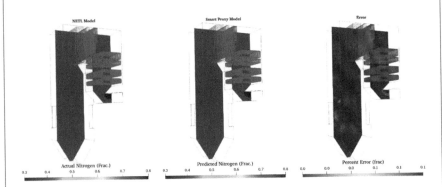

Figure 5.30 Cross section (b) of the nitrogen results – Smart Proxy of the CFD blind validation run number 6.

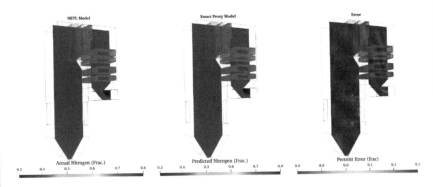

Figure 5.31 Cross section (c) of the nitrogen results – Smart Proxy of the CFD blind validation run number 6.

Error Range (frac)	Simulation Run	Run Type	Number of Cells	Perc. Cells
< 0.02	6	Blind	4,218,832	88.849%
0.02 - 0.04	6	Blind	513,146	10.807%
0.04 - 0.06	6	Blind	13,520	0.285%
0.06 - 0.08	6	Blind	2,465	0.052%
0.08 - 0.1	6	Blind	349	0.007%
> 0.1	6	Blind	16	0.0%

Figure 5.32 Distribution of less than 1% oxygen errors of the CFD blind validation run number 6.

results. For this matter, we trained several models that mainly differ in their number of hidden layers and number of hidden neurons in each hidden layer.

Part of this study was devoted to testing different neural network topologies. This included variation in different numbers of hidden layers as well as hidden neurons in each layer. Results of this approach combined with the updated partitioning (21 simulation runs) are summarized in Table 5.4. Increasing the number of hidden layers (Figure 5.50), various scenarios were tested. Results shown in Figures 5.51–5.57 can be compared with the results shown in Figures 5.45 and 5.46.

What we can gather from this experiment is that increasing the number of layer and neurons in the network structure provides an improvement. At this stage, the highest percentage of cells under 2% error is 66%. While adding more neurons and increasing the complexity of the network is generally helpful, the computational energy required to process data in the network becomes a major constraint on what can be used. Based on our results highlighted in Figure 5.50,

Error Range (frac)	Simulation Run	Run Type	Number of Cells	Perc. Cells
<0.1	6	Blind	1,513,577	31.876%
0.1-0.2	6	Blind	818,769	17.243%
0.2-0.3	6	Blind	771,172	16.241%
0.3-0.4	6	Blind	752,482	15.847%
0.4-0.5	6	Blind	179,073	3.771%
0.5-0.6	6	Blind	80,306	1.691%
0.6-0.7	6	Blind	54,441	1.147%
0.7-0.8	6	Blind	41,799	0.88%
0.8-0.9	6	Blind	33,524	0.706%
> 0.9	6	Blind	456,723	9.619%

Figure 5.33 Distribution of less than 10% oxygen errors of the CFD blind validation run number 6.

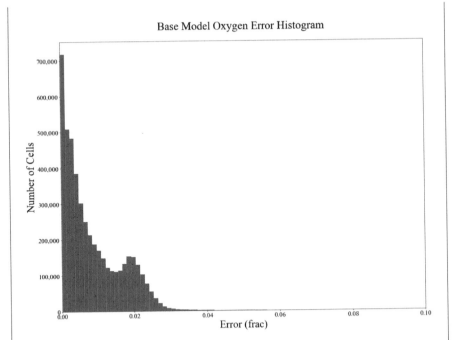

Figure 5.34 Histogram of the oxygen errors of the CFD blind validation run number 6.

the more complex a neural network gets, the longer it takes to train, and thus the less efficient it becomes. The structure that provides the most accurate results in the shortest amount of time is model 3, with 3 hidden layers with 2,000 hidden neurons in each layer. After determining a new network structure, we proceeded with further steps to enhance the model.

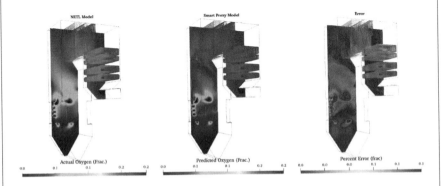

Figure 5.35 Cross section (a) of the oxygen results – Smart Proxy of the CFD blind validation run number 6.

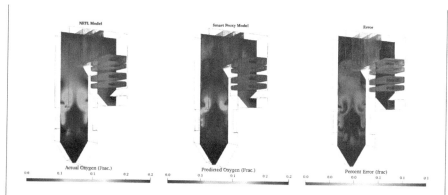

Figure 5.36 Cross section (b) of the oxygen results – Smart Proxy of the CFD blind validation run number 6.

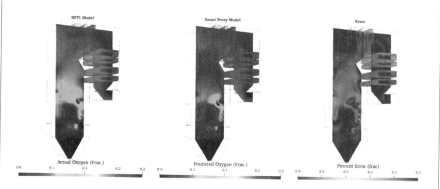

Figure 5.37 Cross section (c) of the oxygen results – Smart Proxy of the CFD blind validation run number 6.

Error Range (frac)	Simulation Run	Run Type	Number of Cells	Perc. Cells
< 0.02	6	Blind	4,654,447	98.023%
0.02 - 0.04	6	Blind	84,711	1.784%
0.04 - 0.06	6	Blind	6,029	0.127%
0.06 - 0.08	6	Blind	1,697	0.036%
0.08 - 0.1	6	Blind	708	0.015%
> 0.1	6	Blind	736	0.016%

Figure 5.38 Distribution of less than 1% pressure errors of the CFD blind validation run number 6.

Error Range (frac)	Simulation Run	Run Type	Number of Cells	Perc. Cells
<0.1	6	Blind	4,569,413	96.232%
0.1-0.2	6	Blind	68,322	1.439%
0.2-0.3	6	Blind	16,638	0.35%
0.3-0.4	6	Blind	13,613	0.287%
0.4-0.5	6	Blind	12,416	0.261%
0.5-0.6	6	Blind	10,043	0.212%
0.6-0.7	6	Blind	5,715	0.12%
0.7-0.8	6	Blind	3,261	0.069%
0.8-0.9	6	Blind	2,696	0.057%
>0.9	6	Blind	46,211	0.973%

Figure 5.39 Distribution of less than 10% pressure errors of the CFD blind validation run number 6.

DATA PARTITIONING (21 SIMULATION RUNS)

The first partitioning approach (14 simulation runs) was revisited and analyzed in more detail. For the second data partitioning attempt, 21 simulation runs were carefully selected based on the minimum and maximum values from both

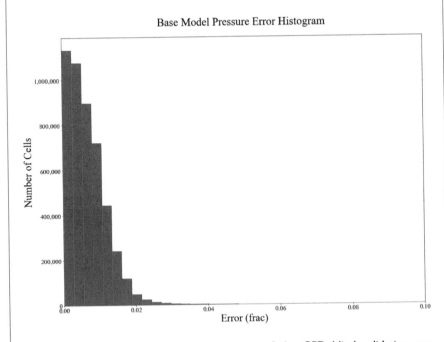

Figure 5.40 Histogram of the pressure errors of the CFD blind validation run number 6.

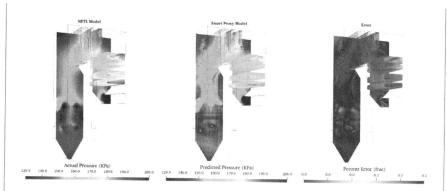

Figure 5.41 Cross section (a) of the pressure results – Smart Proxy of the CFD blind validation run number 6.

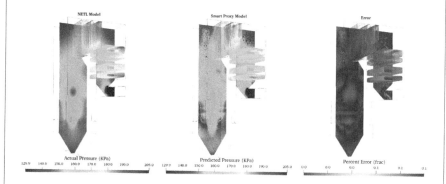

Figure 5.42 Cross section (b) of the pressure results – Smart Proxy of the CFD blind validation run number 6.

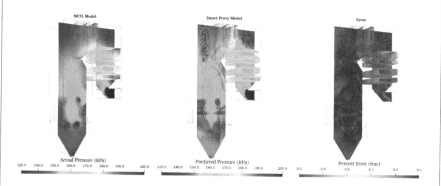

Figure 5.43 Cross section (c) of the pressure results – Smart Proxy of the CFD blind validation run number 6.

Error Range (frac)	Simulation Run	Run Type	Number of Cells	Perc. Cells
< 0.02	6	Blind	2,964,343	62.429%
0.02 - 0.04	6	Blind	1,054,547	22.209%
0.04 - 0.06	6	Blind	326,287	6.872%
0.06 - 0.08	6	Blind	132,197	2.784%
0.08 - 0.1	6	Blind	65,204	1.373%
> 0.1	6	Blind	205,750	4.333%

Figure 5.44 Distribution of less than 1% temperature errors of the CFD blind validation run number 6.

Error Range (frac)	Simulation Run	Run Type	Number of Cells	Perc. Cells
<0.1	6	Blind	4,542,578	95.667%
0.1-0.2	6	Blind	181,356	3.819%
0.2-0.3	6	Blind	14,767	0.311%
0.3-0.4	6	Blind	4,742	0.1%
0.4-0.5	6	Blind	2,344	0.049%
0.5-0.6	6	Blind	1,350	0.028%
0.6-0.7	6	Blind	791	0.017%
0.7-0.8	6	Blind	293	0.006%
0.8-0.9	6	Blind	92	0.002%
> 0.9	6	Blind	15	0.0%

Figure 5.45 Distribution of less than 10% temperature errors of the CFD blind
validation run number 6.

CFD simulation inputs and outputs. Furthermore, an optimum range of values
in-between each of the 21 simulation runs used in the training and calibration
of the Smart Proxy Model was adopted. Figures 5.58–5.60 show the distribution of 21 simulation runs highlighted in blue based on sorted coal, primary, and
secondary flow rates.

During the enhancement process, it was important to quantify any improvement for attempted enhancement steps. Figures 5.58–5.60 also highlight the
improvements seen when modifying the partitioning and adding additional features to the temperature model. Figures 5.58–5.60 represent the results after
deploying the model for simulation runs in the training category. The goal here
was to discern the level of improvement when using 21 training cases instead of 14.

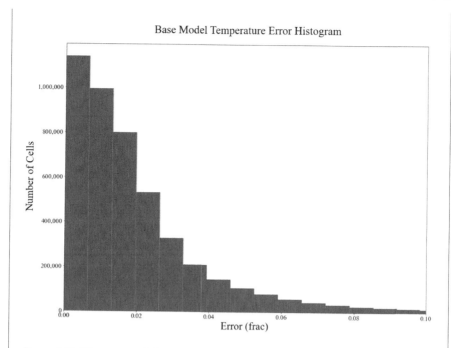

Figure 5.46 Histogram of the temperature errors of the CFD blind validation run number 6.

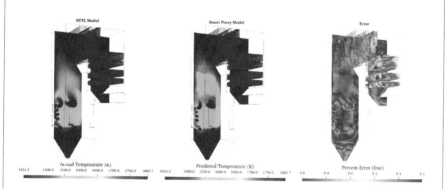

Figure 5.47 Cross section (a) of the temperature results – Smart Proxy of the CFD blind validation run number 6.

The base model in this case (as shown in the bottom of the Figures 5.58–5.60) used 14 training CFD simulation runs. The new partitioning model used 21 training CFD simulation runs. In order to get a representative comparison, the features used and the network structure, as well as the hyper-parameters,

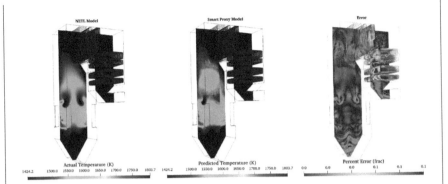

Figure 5.48 Cross section (b) of the temperature results – Smart Proxy of the CFD blind validation run number 6.

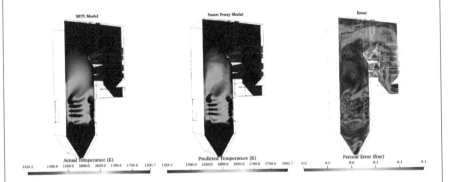

Figure 5.49 Cross section (c) of the temperature results – Smart Proxy of the CFD blind validation run number 6.

Table 5.4 Network structure resource requirements

Model	Number of hidden layers	Number of hidden neurons per layer	Total neurons	Training time per epoch (minutes)	Array size
1	1	1,000	1,000	2	1.32 MB
2	1	10,000	10,000	8.5	12.96 MB
3	3	2,000	6,000	19	94.22 MB
4	3	5,000	15,000	68	578.82 MB
5	3	10,000	30,000	325	2.25 GB
6	5	5,000	25,000	166	1.12 GB
7	5	10,000	50,000	630	4.48 GB

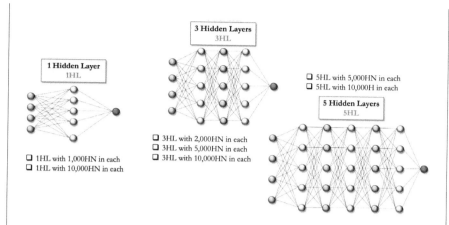

Figure 5.50 Hidden layer experiment visualization.

were identical. In other words, the only difference between the base model and the new partitioning model is the number of simulation runs being included in training.

At the bottom of Figure 5.58, the predictive model's percentage of cells averaged across all training simulation runs with less than 2%, 4%, and 6% error are shown. To put it simply, the base model used 14 simulation runs and when we apply a filter to only show cells whose prediction is within 2% of the actual CFD model, that selection accounts for 85.5% of the total cells for the base model and 84.9% of the same total cells for the new partitioning model.

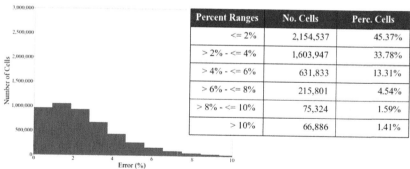

Percent Ranges	No. Cells	Perc. Cells
<= 2%	2,154,537	45.37%
> 2% - <= 4%	1,603,947	33.78%
> 4% - <= 6%	631,833	13.31%
> 6% - <= 8%	215,801	4.54%
> 8% - <= 10%	75,324	1.59%
> 10%	66,886	1.41%

Figure 5.51 Distribution of less than 10% temperature errors and histogram of the temperature errors of the CFD blind validation run number 6 for model 1 including 1 hidden layer and 1,000 hidden neurons.

1 Hidden Layer x 10,000 Hidden Neurons

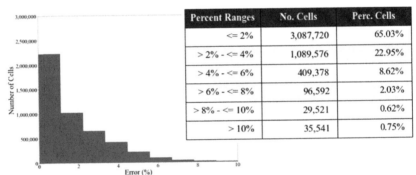

Percent Ranges	No. Cells	Perc. Cells
<= 2%	2,729,779	57.49%
> 2% - <= 4%	1,253,466	26.4%
> 4% - <= 6%	533,200	11.23%
> 6% - <= 8%	134,457	2.83%
> 8% - <= 10%	43,334	0.91%
> 10%	54,092	1.14%

Figure 5.52 Distribution of less than 10% temperature errors and histogram of the temperature errors of the CFD blind validation run number 6 for model 2 including 1 hidden layer and 10,000 hidden neurons.

3 Hidden Layer x 2,000 Hidden Neurons

Percent Ranges	No. Cells	Perc. Cells
<= 2%	3,087,720	65.03%
> 2% - <= 4%	1,089,576	22.95%
> 4% - <= 6%	409,378	8.62%
> 6% - <= 8%	96,592	2.03%
> 8% - <= 10%	29,521	0.62%
> 10%	35,541	0.75%

Figure 5.53 Distribution of less than 10% temperature errors and histogram of the temperature errors of the CFD blind validation run number 6 for model 3 including 3 hidden layers and 2,000 hidden neurons.

At the bottom of Figure 5.59, another perspective for evaluating the level of improvement is shown. This table looks at the results of all the simulation run predictions and simply returns the error of the lowest performing simulation run. Here we see that for cells under 2% error, the base model at its worse will have 83.9% of the total cells within 2% of the CFD model prediction.

Bottom of Figure 5.60 shows a perspective on the other side of the spectrum. This highlights the best or most accurate simulation run prediction. Here we see that base model at its best can provide a prediction where 86% of the cells are

3 Hidden Layer x 5,000 Hidden Neurons

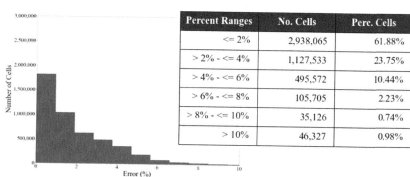

Percent Ranges	No. Cells	Perc. Cells
<= 2%	2,938,065	61.88%
> 2% - <= 4%	1,127,533	23.75%
> 4% - <= 6%	495,572	10.44%
> 6% - <= 8%	105,705	2.23%
> 8% - <= 10%	35,126	0.74%
> 10%	46,327	0.98%

Figure 5.54 Distribution of less than 10% temperature errors and histogram of the temperature errors of the CFD blind validation run number 6 for model 4 including 3 hidden layers and 5,000 hidden neurons.

3 Hidden Layer x 10,000 Hidden Neurons

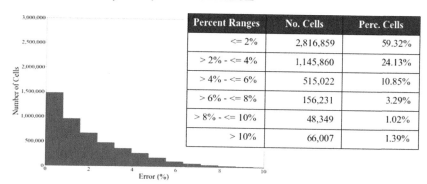

Percent Ranges	No. Cells	Perc. Cells
<= 2%	2,816,859	59.32%
> 2% - <= 4%	1,145,860	24.13%
> 4% - <= 6%	515,022	10.85%
> 6% - <= 8%	156,231	3.29%
> 8% - <= 10%	48,349	1.02%
> 10%	66,007	1.39%

Figure 5.55 Distribution of less than 10% temperature errors and histogram of the temperature errors of the CFD blind validation run number 6 for model 5 including 3 hidden layers and 10,000 hidden neurons.

within 2%, or 98% accurate, of the actual CFD model. What we can gather here is that the performance of the training cases appears similar, but it is more practical to analyze the results of the blind simulation runs to gauge how the model may perform in the real world. While the training performance appears similar, there is significant disparity when studying the blind performance.

Figure 5.61 illustrates that the average error for blind simulation runs, simulation runs whose inputs have not been exposed to the network, is roughly 10% better. It can be seen that the new partitioning model, on average, will provide more cells whose prediction is closer to the actual CFD model.

5 Hidden Layer x 5,000 Hidden Neurons

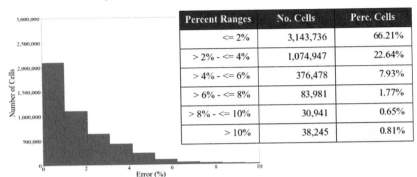

Percent Ranges	No. Cells	Perc. Cells
<= 2%	3,143,736	66.21%
> 2% - <= 4%	1,074,947	22.64%
> 4% - <= 6%	376,478	7.93%
> 6% - <= 8%	83,981	1.77%
> 8% - <= 10%	30,941	0.65%
> 10%	38,245	0.81%

Figure 5.56 Distribution of less than 10% temperature errors and histogram of the temperature errors of the CFD blind validation run number 6 for model 6 including 5 hidden layers and 5,000 hidden neurons.

5 Hidden Layer x 10,000 Hidden Neurons

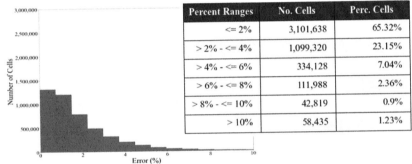

Percent Ranges	No. Cells	Perc. Cells
<= 2%	3,101,638	65.32%
> 2% - <= 4%	1,099,320	23.15%
> 4% - <= 6%	334,128	7.04%
> 6% - <= 8%	111,988	2.36%
> 8% - <= 10%	42,819	0.9%
> 10%	58,435	1.23%

Figure 5.57 Distribution of less than 10% temperature errors and histogram of the temperature errors of the CFD blind validation run number 6 for model 7 including 5 hidden layers and 10,000 hidden neurons.

FEATURE GENERATION

As was mentioned in Chapter 4, one of the most important parts of the Smart Proxy Modeling development is domain expertise that would provide feature generation, which enhances the way of teaching physical technology that has been numerically simulated. In this section, feature generation for the CFD Smart Proxy Model for this particular problem is being covered. This feature generation effort reinforces the Smart Proxy Model's relationship between inputs and outputs. The following features are mostly related to the boiler's geometry.

Simulation Run	Coal (Sorted)	Primary	Secondary
25	0.65	2.32	2.91
16	0.68	3.09	4.16
3	0.71	2.70	5.41
15	0.74	3.47	3.33
31	0.78	2.51	4.58
1	0.81	3.28	5.83
18	0.84	2.90	3.74
21	0.88	3.66	5.00
22	0.91	2.42	6.25
19	0.94	3.18	3.05
5	0.97	2.80	4.30
20	1.01	3.56	5.55
27	1.04	2.61	3.47
6	1.07	3.37	4.72
10	1.11	2.99	5.97
11	1.14	3.75	3.88
17	1.17	2.37	5.13
26	1.21	3.13	6.39
9	1.24	2.75	3.19
23	1.27	3.51	4.44
7	1.30	2.56	5.69
24	1.34	3.32	3.60
8	1.37	2.94	4.86
13	1.40	3.70	6.11
32	1.44	2.47	4.02
30	1.47	3.23	5.27
12	1.50	2.85	6.53
4	1.53	3.61	2.95
28	1.57	2.66	4.21
14	1.60	3.42	5.46
29	1.63	3.04	3.37
2	1.67	3.80	4.62

TRAINING COMPARISON		Percentage of cells within given error threshold averaged across all simulation runs used in TRAINING		
Model	Number of Input Features	Cells under 2% Error	Cells under 4% Error	Cells under 6% Error
Base Model	84	85.5	93.8	96.5
New Partitioning	84	84.9	93.9	96.6

Figure 5.58 The number of CFD simulation runs used for the new Smart Proxy Model development shown in blue background, including training comparisons. Training CFD runs number 1.

OUTLET DUCT DISTANCES

Like the distances calculated from each cell to both the wall and inlet boundary conditions, distances from each cell to a center point at the outlet duct were added to the training dataset. Figure 5.62 roughly shows the different locations

Simulation Run	Primary (Sorted)	Coal	Secondary
25	2.32	0.65	2.91
17	2.37	1.17	5.13
22	2.42	0.91	6.25
32	2.47	1.44	4.02
31	2.51	0.78	4.58
7	2.56	1.30	5.69
27	2.61	1.04	3.47
28	2.66	1.57	4.21
3	2.70	0.71	5.41
9	2.75	1.24	3.19
5	2.80	0.97	4.30
12	2.85	1.50	6.53
18	2.90	0.84	3.74
8	2.94	1.37	4.86
10	2.99	1.11	5.97
29	3.04	1.63	3.37
16	3.09	0.68	4.16
26	3.13	1.21	6.39
19	3.18	0.94	3.05
30	3.23	1.47	5.27
1	3.28	0.81	5.83
24	3.32	1.34	3.60
6	3.37	1.07	4.72
14	3.42	1.60	5.46
15	3.47	0.74	3.33
23	3.51	1.27	4.44
20	3.56	1.01	5.55
4	3.61	1.53	2.95
21	3.66	0.88	5.00
13	3.70	1.40	6.11
11	3.75	1.14	3.88
2	3.80	1.67	4.62

TRAINING COMPARISON		The lowest percentage of cells within given error threshold for all simulaion runs used in TRAINING		
Model	Number of Input Features	Cells under 2% Error	Cells under 4% Error	Cells under 6% Error
Base Model	84	83.9	91.9	94.6
New Partitioning	84	79.3	90.2	93.6

Figure 5.59 The number of CFD simulation runs used for the new Smart Proxy Model development shown in blue background, including training comparisons. Training CFD runs number 2.

of a cell and the trajectories considered when calculating the distance to the outlet of the boiler.

NEAREST PLATEN DISTANCE

Each platen had a function in the boiler, whether it was an economizer that received the supplied water into the boiler, or a superheater that further heated

Simulation Run	Secondary (Sorted)	Coal	Primary
25	2.91	0.65	2.32
4	2.95	1.53	3.61
19	3.05	0.94	3.18
9	3.19	1.24	2.75
15	3.33	0.74	3.47
29	3.37	1.63	3.04
27	3.47	1.04	2.61
24	3.60	1.34	3.32
18	3.74	0.84	2.90
11	3.88	1.14	3.75
32	4.02	1.44	2.47
16	4.16	0.68	3.09
28	4.21	1.57	2.66
5	4.30	0.97	2.80
23	4.44	1.27	3.51
31	4.58	0.78	2.51
2	4.62	1.67	3.80
6	4.72	1.07	3.37
8	4.86	1.37	2.94
21	5.00	0.88	3.66
17	5.13	1.17	2.37
30	5.27	1.47	3.23
3	5.41	0.71	2.70
14	5.46	1.60	3.42
20	5.55	1.01	3.56
7	5.69	1.30	2.56
1	5.83	0.81	3.28
10	5.97	1.11	2.99
13	6.11	1.40	3.70
22	6.25	0.91	2.42
26	6.39	1.21	3.13
12	6.53	1.50	2.85

TRAINING COMPARISON		The highest percentage of cells within given error threshold for all simulaion runs used in TRAINING		
Model	Number of Input Features	Cells under 2% Error	Cells under 4% Error	Cells under 6% Error
Base Model	84	85.97	95.07	97.74
New Partitioning	84	88.52	95.96	98.01

Figure 5.60 The number of CFD simulation runs used for the new Smart Proxy Model development shown in blue background, including training comparisons. Training CFD runs number 3.

the wet steam, or a reheater that reheated the recycled steam coming from the turbines. As a matter of fact, the location of each cell with respect to each platen was important and communicated to the neural network by identifying the distance to the closest superheater, reheater, or economizer.

Figure 5.63 provides a variety of scenarios to determine the distance to the nearest platen. In each scenario, there are two arrows (dashed red and solid gray)

BLIND COMPARISON		Percentage of cells within given error threshold averaged across all simulation runs used in BLIND		
Model	Number of Input Features	Cells under 2% Error	Cells under 4% Error	Cells under 6% Error
Base Model	84	52.79	75.5	85.5
New Partitioning	84	66.1	86.4	93.8

BLIND COMPARISON		The lowest percentage of cells within given error threshold for all simulaion runs used in BLIND		
Model	Number of Input Features	Cells under 2% Error	Cells under 4% Error	Cells under 6% Error
Base Model	84	4.8	9.8	15.8
New Partitioning	84	35.2	68.0	84.3

BLIND COMPARISON		The highest percentage of cells within given error threshold for all simulaion runs used in BLIND		
Model	Number of Input Features	Cells under 2% Error	Cells under 4% Error	Cells under 6% Error
Base Model	84	78.4	92.9	96.6
New Partitioning	84	84.7	94.5	97.5

Figure 5.61 Results of the blind validation using 21 CFD simulation runs used for the new Smart Proxy Model development.

that represent the nearest platens. However, the dashed red arrow is the primary platen closer to the cell.

REGIONS

A one-hot encoding approach was implemented to provide information regarding the seven numerical simulation regions. This approach focused on identifying groups of cells with similar sizes and volumes, which provided more geometrical information about the boiler. As mentioned in the description of the simulation

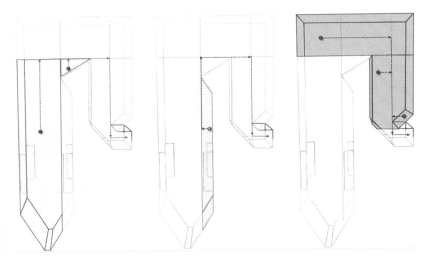

Figure 5.62 Feature generation: Outlet distances.

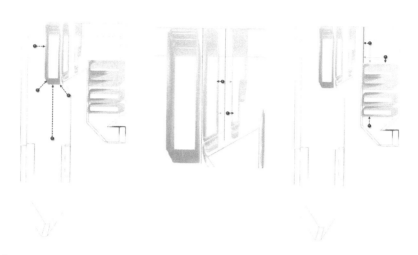

Figure 5.63 Feature generation: Nearest platen.

| | Feature Generation | | | | | | |
| | Regions (7) | | | | | | |
Cell ID	Hopper	Hopper Above	Windbox	Windbox Above	Arch	Rear Pass	Rear Pass
4277586	1	0	0	0	0	0	0
2849239	0	1	0	0	0	0	0
1414076	0	0	1	0	0	0	0
4611868	0	0	0	1	0	0	0
2556442	0	0	0	0	1	0	0
2930209	0	0	0	0	0	1	0
4448953	0	0	0	0	0	0	1

Figure 5.64 Feature generation: Numerical simulation regions.

model, these regions were created due to sudden geometry changes as well as highly complex reactions (Figure 5.64).

PLATEN COUNT

A relationship between the inlet and outlet boundary conditions was established by providing information of the platens (superheaters, reheaters, and economizers) encountered from the inlet to the outlet and vice versa. This approach was implemented in a dynamic fashion.

For example, Cell ID 10 in Figure 5.65, is located at the "Superheater Horizontal Lower". The number of platens the hot gas would have traversed from the moment of combustion at the inlet to the current cell location would have been six, starting with the superheater, followed by the reheater front, reheater rear, superheater vertical, superheater horizontal upper, and ending

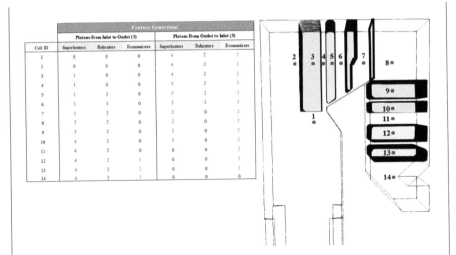

	Feature Generation					
	Platens From Inlet to Outlet (3)			Platens From Outlet to Inlet (3)		
Cell ID	Superheaters	Reheaters	Economizers	Superheaters	Reheaters	Economizers
1	0	0	0	4	2	2
2	0	0	0	4	2	2
3	1	0	0	4	2	2
4	1	0	0	3	2	2
5	1	1	0	3	2	2
6	1	1	0	3	1	2
7	1	2	0	3	0	2
8	2	2	0	2	0	2
9	3	2	0	2	0	2
10	4	2	0	1	0	2
11	4	2	0	0	0	2
12	4	2	1	0	0	2
13	4	2	2	0	0	1
14	4	2	2	0	0	0

Figure 5.65 Feature generation: Platen count.

at the current superheater horizontal lower platen. This is summarized in the first three columns of the table, where the number of superheaters, reheaters, and economizers is four, two, and zero, respectively. Similarly, the number of platens traversed from the outlet to the inlet would have been three; starting with the economizer lower, followed by the economizer upper, and finalizing at the superheater horizontal lower.

TURNS

Another dynamic approach was implemented to provide a sense of direction and space in the boiler by adding information regarding the number of turns that a cell would have made in order to reach the inlet and the outlet. Figure 5.66 shows different cell locations and their journey toward the outlet of the system. The path switches direction based on the model's geometry and thus a "turn" is made in efforts to reach the outlet. The red circles represent the total number of turns made by a particular cell. Figure 5.67 presents a similar scenario. However, the path is in the direction toward the inlet.

We can also see that the base model, although trained similarly, was not able to generalize the physics and mechanisms to reliably predict simulation run outputs. While the worse performing new partitioning model simulation run only managed to get 35.2% of the cells to under 2% error, it still managed to get 84.3% of the cells under 6% error instead of 15.8%. This shows that using the new partitioning is arguably at least five times more reliable (Figure 5.68).

Based on the discussion about how to interpret the information shown in Figure 5.68, a similar conclusion can be drawn for the use of additional features;

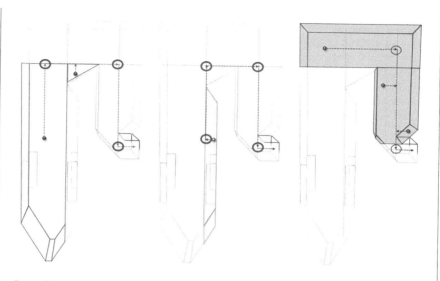

Figure 5.66 Feature generation: Turns towards the outlet.

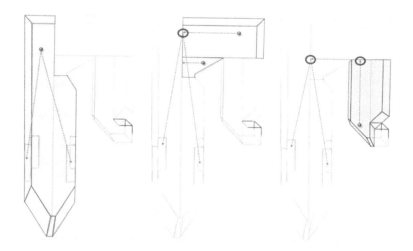

Figure 5.67 Feature generation: Turns to inlet.

however, the impact is not so substantial. An example of the improvement is shown in Figure 5.69.

In Figure 5.69, the location shown is a horizontal cross section of the combustion occuring with the injectors in the corners being viewed from above. Here we can see that the actual CFD model (labeled NETL model on the left side) is more similar to the improved temperature model versus the base temperature model. The colors blue, green, and yellow represent key temperatures within

TRAINING COMPARISON		Percentage of cells within given error threshold averaged across all simulation runs used in TRAINING		
Model	Number of Input Features	Cells under 2% Error	Cells under 4% Error	Cells under 6% Error
Base Model	84	85.5	93.8	96.5
New Features	101	89.1	94.5	96.1

TRAINING COMPARISON		The lowest percentage of cells within given error threshold for all simulaion runs used in TRAINING		
Model	Number of Input Features	Cells under 2% Error	Cells under 4% Error	Cells under 6% Error
Base Model	84	83.9	91.9	94.6
New Features	101	87.7	93.2	95.1

TRAINING COMPARISON		The highest percentage of cells within given error threshold for all simulaion runs used in TRAINING		
Model	Number of Input Features	Cells under 2% Error	Cells under 4% Error	Cells under 6% Error
Base Model	84	85.97	95.07	97.74
New Features	101	90.44	95.45	96.85

BLIND COMPARISON		Percentage of cells within given error threshold averaged across all simulation runs used in BLIND		
Model	Number of Input Features	Cells under 2% Error	Cells under 4% Error	Cells under 6% Error
Base Model	84	52.79	75.5	85.5
New Features	101	55.5	78.3	88.1

BLIND COMPARISON		The lowest percentage of cells within given error threshold for all simulaion runs used in BLIND		
Model	Number of Input Features	Cells under 2% Error	Cells under 4% Error	Cells under 6% Error
Base Model	84	4.8	9.8	15.8
New Features	101	34.4	55.0	64.1

BLIND COMPARISON		The highest percentage of cells within given error threshold for all simulaion runs used in BLIND		
Model	Number of Input Features	Cells under 2% Error	Cells under 4% Error	Cells under 6% Error
Base Model	84	78.4	92.9	96.6
New Features	101	79.5	92.4	95.6

Figure 5.68 Results of the training and blind validation using 21 CFD simulation runs and including new feature generation.

the flame. The base model (on the right side) shows a hotter, red color, temperature just outside the rear left injector compared to the actual model. Also notice the behavior of the center of the flame; the mixing and combustion are captured more in the improved model.

SEQUENTIAL MODELING

One of the main challenges faced in this project was predicting gas concentrations, pressure, and temperature distributions data that was to a great extent related to the boiler's geometry – distances, adjacent cells, number of turns, number of platens, etc. Sequential modeling was a potential solution to this problem in which predictions from other attributes of interest could be used as an input to a model under development.

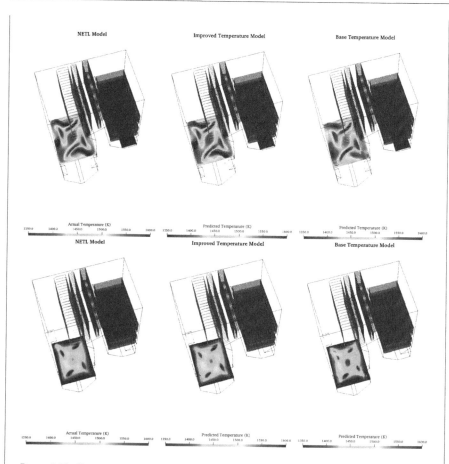

Figure 5.69 Cross-sectional comparison between initial model (base), improved model, and actual model.

This approach was tested by deploying pressure, CO_2, N_2, and O_2 models and using the predicted data to train and deploy the temperature model (Figure 5.70). The temperature model was trained with the actual solution output data of pressure, CO_2, N_2, and O_2, and calibrated based on the predicted data.

FINAL CFD SMART PROXY RESULTS

In this section, the final results of the Tri-State boiler's CFD Smart Proxy Model is presented. The results show the complete distribution of five different parameters in the Tri-State boiler that are importantly modeled in the CFD. These five parameters are pressure, carbon dioxide, oxygen, nitrogen, and temperature. Furthermore, the results that are presented in this section

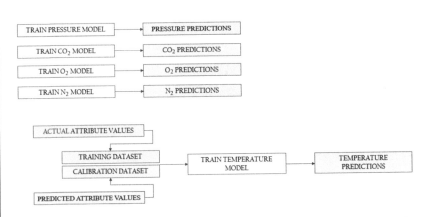

Figure 5.70 CO$_2$, N$_2$, and O$_2$ models and using the predicted data to train and deploy the temperature model.

are from two CFD simulation runs that were not used during the develoment of the Tri-State boiler's CFD Smart Proxy Model. These two CSD simulation runs that are referred to as blind validations are named as CFD simulation run number 5 and CFD simulation run number 6. These results are shown in Figures 5.71–5.120.

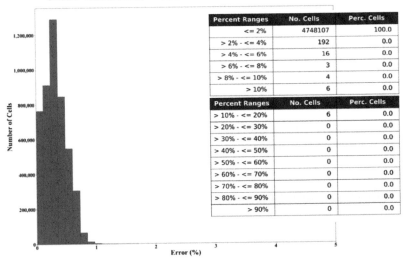

Smart Proxy Model: **Pressure**

Blind Validation – CFD Simulation Run #5

Percent Ranges	No. Cells	Perc. Cells
<= 2%	4748107	100.0
> 2% - <= 4%	192	0.0
> 4% - <= 6%	16	0.0
> 6% - <= 8%	3	0.0
> 8% - <= 10%	4	0.0
> 10%	6	0.0

Percent Ranges	No. Cells	Perc. Cells
> 10% - <= 20%	6	0.0
> 20% - <= 30%	0	0.0
> 30% - <= 40%	0	0.0
> 40% - <= 50%	0	0.0
> 50% - <= 60%	0	0.0
> 60% - <= 70%	0	0.0
> 70% - <= 80%	0	0.0
> 80% - <= 90%	0	0.0
> 90%	0	0.0

Figure 5.71 Results of pressure distribution for blind validation CFD simulation run number 5 – Histogram.

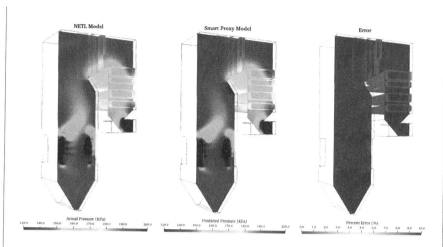

Figure 5.72 Pressure results for simulation run number 5 – Cross section number 1.

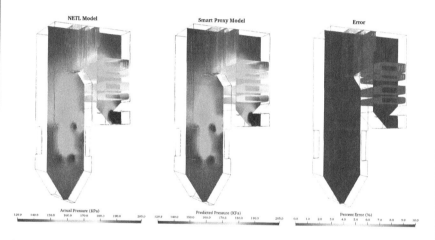

Figure 5.73 Pressure results for simulation run number 5 – Cross section number 2.

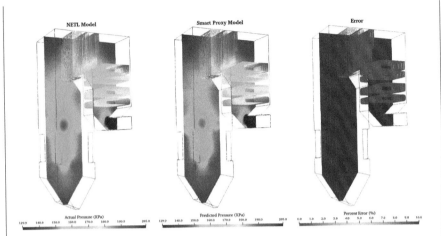

Figure 5.74 Pressure results for simulation run number 5 – Cross section number 3.

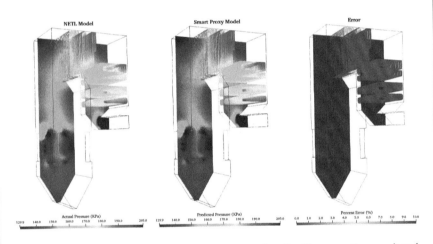

Figure 5.75 Pressure results for simulation run number 5 – Cross section number 4.

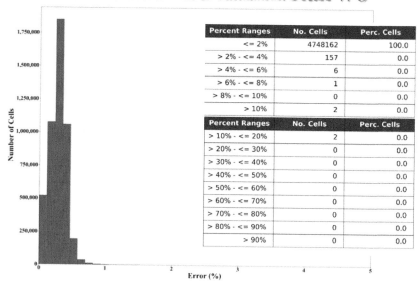

Smart Proxy Model: **Pressure**
Blind Validation – CFD Simulation Run #6

Percent Ranges	No. Cells	Perc. Cells
<= 2%	4748162	100.0
> 2% - <= 4%	157	0.0
> 4% - <= 6%	6	0.0
> 6% - <= 8%	1	0.0
> 8% - <= 10%	0	0.0
> 10%	2	0.0

Percent Ranges	No. Cells	Perc. Cells
> 10% - <= 20%	2	0.0
> 20% - <= 30%	0	0.0
> 30% - <= 40%	0	0.0
> 40% - <= 50%	0	0.0
> 50% - <= 60%	0	0.0
> 60% - <= 70%	0	0.0
> 70% - <= 80%	0	0.0
> 80% - <= 90%	0	0.0
> 90%	0	0.0

Figure 5.76 Results of pressure distribution for blind validation CFD simulation run number 6 – Histogram.

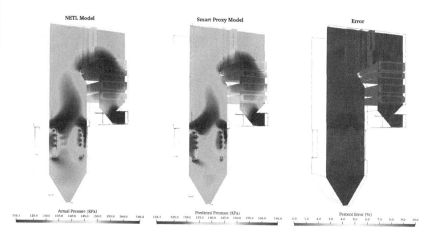

Figure 5.77 Pressure results for simulation run number 6 – Cross section number 1.

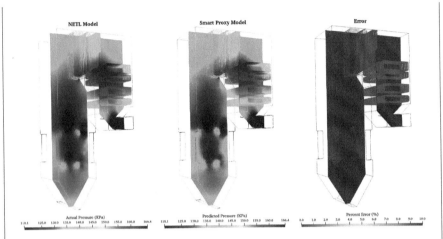

Figure 5.78 Pressure results for simulation run number 6 – Cross section number 2.

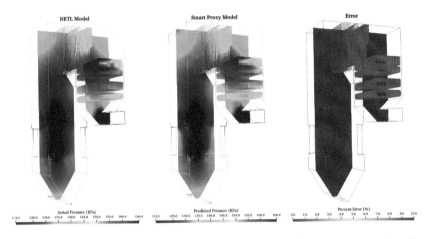

Figure 5.79 Pressure results for simulation run number 6 – Cross section number 3.

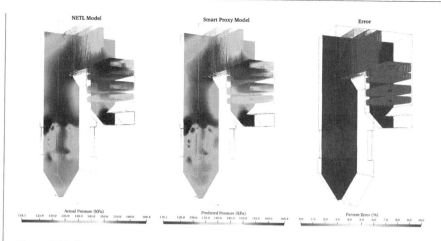

Figure 5.80 Pressure results for simulation run number 6 – Cross section number 4.

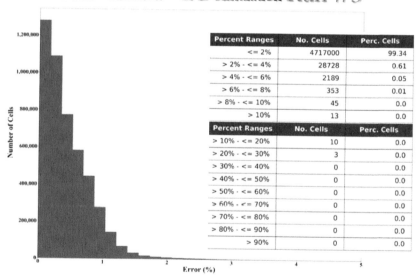

Smart Proxy Model: **Carbon Dioxide**
Blind Validation – CFD Simulation Run #5

Percent Ranges	No. Cells	Perc. Cells
<= 2%	4717000	99.34
> 2% - <= 4%	28728	0.61
> 4% - <= 6%	2189	0.05
> 6% - <= 8%	353	0.01
> 8% - <= 10%	45	0.0
> 10%	13	0.0
Percent Ranges	No. Cells	Perc. Cells
> 10% - <= 20%	10	0.0
> 20% - <= 30%	3	0.0
> 30% - <= 40%	0	0.0
> 40% - <= 50%	0	0.0
> 50% - <= 60%	0	0.0
> 60% - <= 70%	0	0.0
> 70% - <= 80%	0	0.0
> 80% - <= 90%	0	0.0
> 90%	0	0.0

Figure 5.81 Results of carbon dioxide distribution for blind validation CFD simulation run number 5 – Histogram.

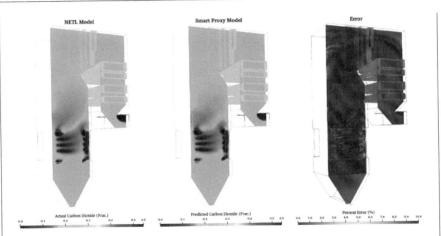

Figure 5.82 Carbon dioxide results for simulation run number 5 – Cross section number 1.

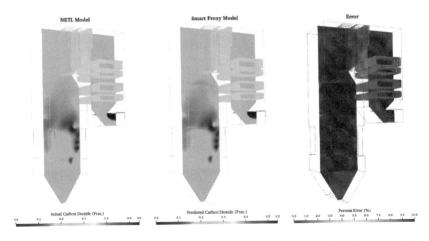

Figure 5.83 Carbon dioxide results for simulation run number 5 – Cross section number 2.

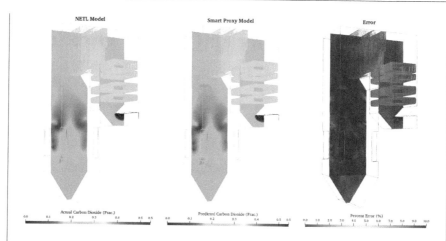

Figure 5.84 Carbon dioxide results for simulation run number 5 – Cross section number 3.

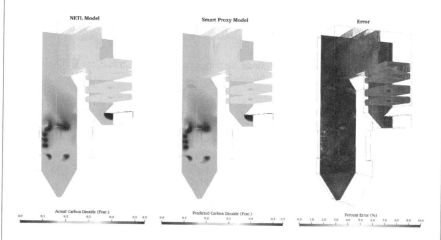

Figure 5.85 Carbon dioxide results for simulation run number 5 – Cross section number 4.

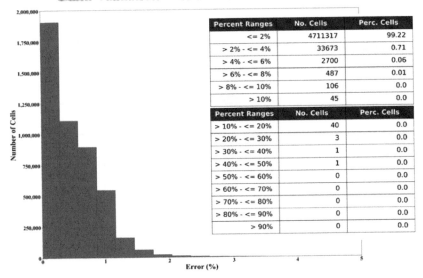

Smart Proxy Model: Carbon Dioxide
Blind Validation – CFD Simulation Run #6

Percent Ranges	No. Cells	Perc. Cells
<= 2%	4711317	99.22
> 2% - <= 4%	33673	0.71
> 4% - <= 6%	2700	0.06
> 6% - <= 8%	487	0.01
> 8% - <= 10%	106	0.0
> 10%	45	0.0

Percent Ranges	No. Cells	Perc. Cells
> 10% - <= 20%	40	0.0
> 20% - <= 30%	3	0.0
> 30% - <= 40%	1	0.0
> 40% - <= 50%	1	0.0
> 50% - <= 60%	0	0.0
> 60% - <= 70%	0	0.0
> 70% - <= 80%	0	0.0
> 80% - <= 90%	0	0.0
> 90%	0	0.0

Figure 5.86 Results of carbon dioxide distribution for blind validation CFD simulation run number 6 – Histogram.

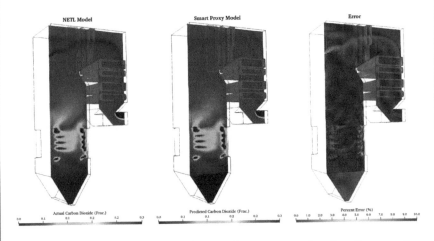

Figure 5.87 Carbon dioxide results for simulation run number 6 – Cross section number 1.

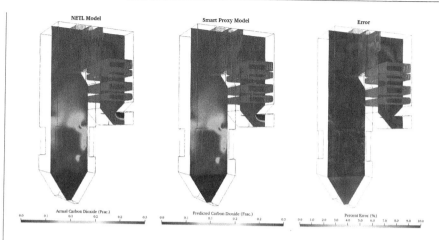

Figure 5.88 Carbon dioxide results for simulation run number 6 – Cross section number 2.

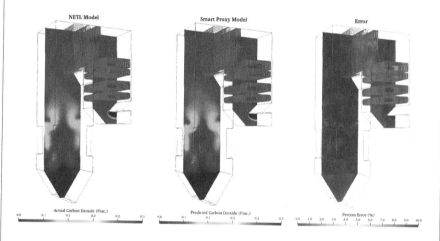

Figure 5.89 Carbon dioxide results for simulation run number 6 – Cross section number 3.

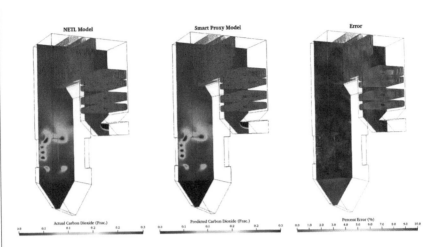

Figure 5.90 Carbon dioxide results for simulation run number 6 – Cross section number 4.

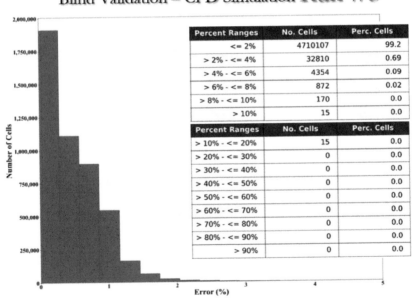

Figure 5.91 Results of oxygen distribution for blind validation CFD simulation run number 5 – Histogram.

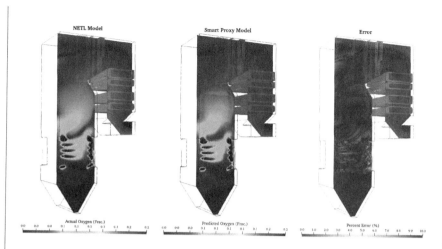

Figure 5.92   Oxygen results for simulation run number 5 – Cross section number 1.

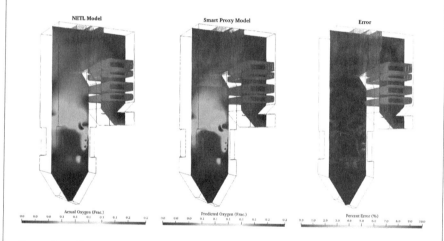

Figure 5.93   Oxygen results for simulation run number 5 – Cross section number 2.

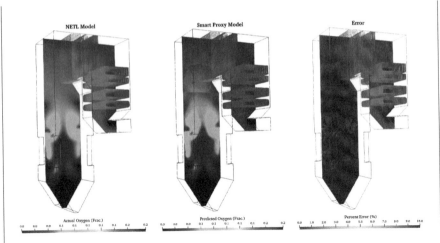

Figure 5.94 Oxygen results for simulation run number 5 – Cross section number 3.

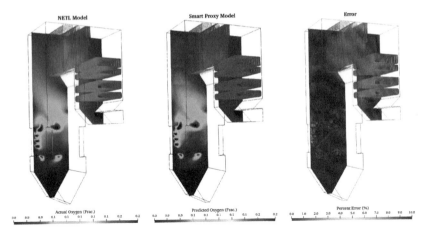

Figure 5.95 Oxygen results for simulation run number 5 – Cross section number 4.

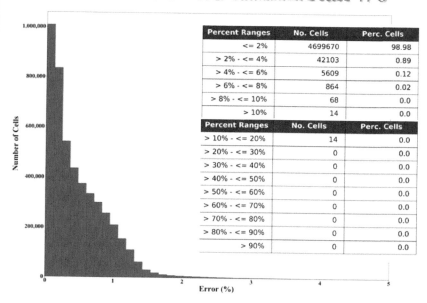

Smart Proxy Model: **Oxygen**
Blind Validation – CFD Simulation Run #6

Percent Ranges	No. Cells	Perc. Cells
<= 2%	4699670	98.98
> 2% - <= 4%	42103	0.89
> 4% - <= 6%	5609	0.12
> 6% - <= 8%	864	0.02
> 8% - <= 10%	68	0.0
> 10%	14	0.0
Percent Ranges	**No. Cells**	**Perc. Cells**
> 10% - <= 20%	14	0.0
> 20% - <= 30%	0	0.0
> 30% - <= 40%	0	0.0
> 40% - <= 50%	0	0.0
> 50% - <= 60%	0	0.0
> 60% - <= 70%	0	0.0
> 70% - <= 80%	0	0.0
> 80% - <= 90%	0	0.0
> 90%	0	0.0

Figure 5.96 Results of oxygen distribution for blind validation CFD simulation run number 6 – Histogram.

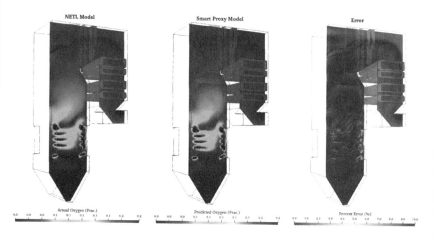

Figure 5.97 Oxygen results for simulation run number 6 – Cross section number 1.

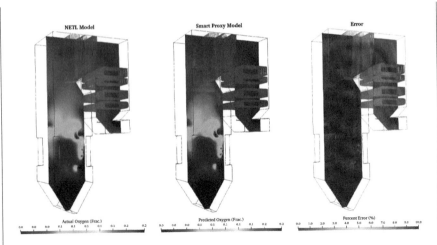

Figure 5.98 Oxygen results for simulation run number 6 – Cross section number 2.

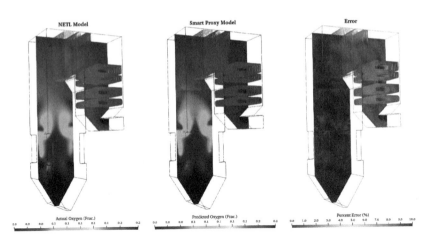

Figure 5.99 Oxygen results for simulation run number 6 – Cross section number 3.

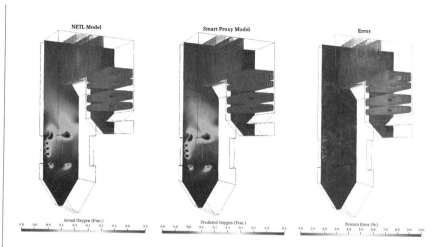

Figure 5.100 Oxygen results for simulation run number 6 – Cross section number 4.

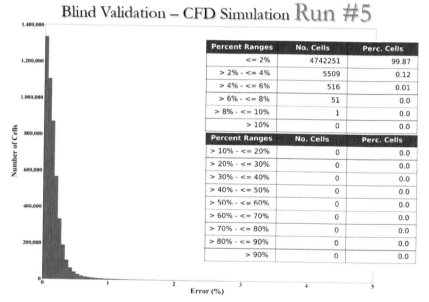

Figure 5.101 Results of nitrogen distribution for blind validation CFD simulation run number 5 – Histogram.

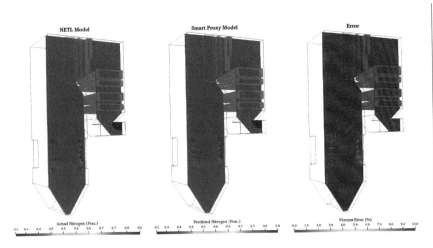

Figure 5.102 Nitrogen results for simulation run number 5 – Cross section number 1.

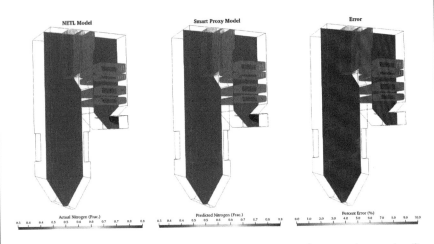

Figure 5.103 Nitrogen results for simulation run number 5 – Cross section number 2.

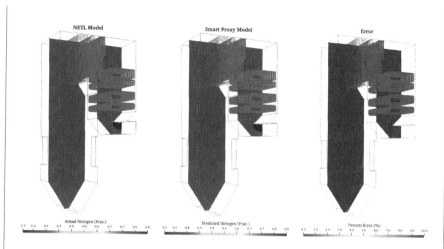

Figure 5.104 Nitrogen results for simulation run number 5 – Cross section number 3.

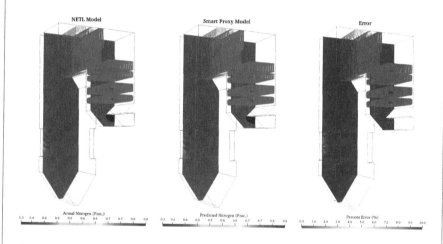

Figure 5.105 Nitrogen results for simulation run number 5 – Cross section number 4.

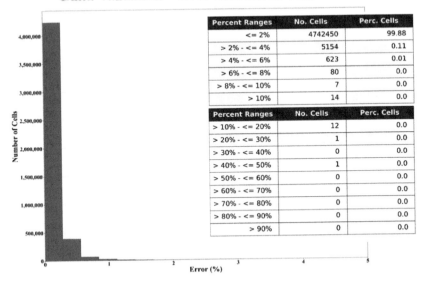

Smart Proxy Model: **Nitrogen**
Blind Validation – CFD Simulation Run #6

Percent Ranges	No. Cells	Perc. Cells
<= 2%	4742450	99.88
> 2% - <= 4%	5154	0.11
> 4% - <= 6%	623	0.01
> 6% - <= 8%	80	0.0
> 8% - <= 10%	7	0.0
> 10%	14	0.0

Percent Ranges	No. Cells	Perc. Cells
> 10% - <= 20%	12	0.0
> 20% - <= 30%	1	0.0
> 30% - <= 40%	0	0.0
> 40% - <= 50%	1	0.0
> 50% - <= 60%	0	0.0
> 60% - <= 70%	0	0.0
> 70% - <= 80%	0	0.0
> 80% - <= 90%	0	0.0
> 90%	0	0.0

Figure 5.106 Results of nitrogen distribution for blind validation CFD simulation run number 6 – Histogram.

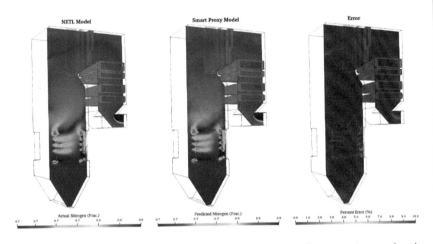

Figure 5.107 Nitrogen results for simulation run number 6 – Cross section number 1.

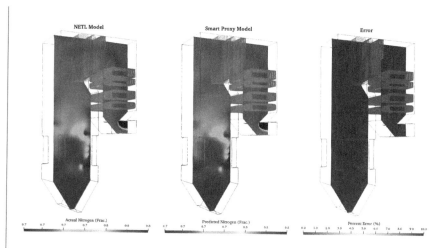

Figure 5.108 Nitrogen results for simulation run number 6 – Cross section number 2.

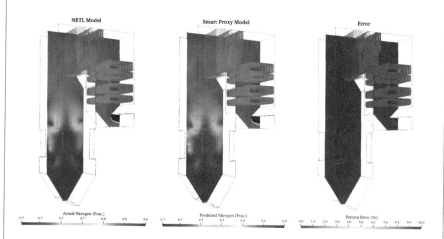

Figure 5.109 Nitrogen results for simulation run number 6 – Cross section number 3.

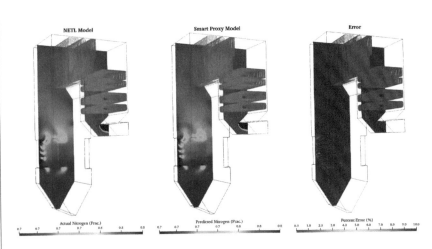

Figure 5.110 Nitrogen results for simulation run number 6 – Cross section number 4.

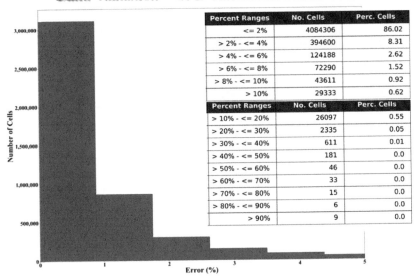

Percent Ranges	No. Cells	Perc. Cells
<= 2%	4084306	86.02
> 2% - <= 4%	394600	8.31
> 4% - <= 6%	124188	2.62
> 6% - <= 8%	72290	1.52
> 8% - <= 10%	43611	0.92
> 10%	29333	0.62
Percent Ranges	**No. Cells**	**Perc. Cells**
> 10% - <= 20%	26097	0.55
> 20% - <= 30%	2335	0.05
> 30% - <= 40%	611	0.01
> 40% - <= 50%	181	0.0
> 50% - <= 60%	46	0.0
> 60% - <= 70%	33	0.0
> 70% - <= 80%	15	0.0
> 80% - <= 90%	6	0.0
> 90%	9	0.0

Figure 5.111 Results of temperature distribution for blind validation CFD simulation run number 5 – Histogram.

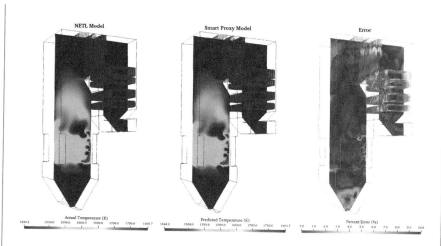

Figure 5.112 Temperature results for simulation run number 5 – Cross section number 1.

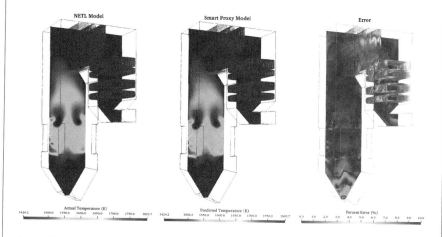

Figure 5.113 Temperature results for simulation run number 5 – Cross section number 2.

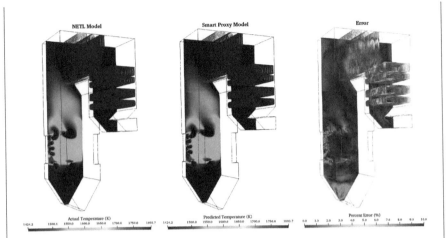

Figure 5.114 Temperature results for simulation run number 5 – Cross section number 3.

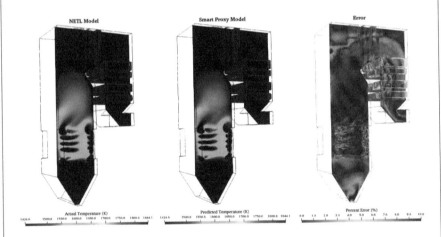

Figure 5.115 Temperature results for simulation run number 5 – Cross section number 4.

Smart Proxy Model: **Temperature**

Blind Validation – CFD Simulation Run #6

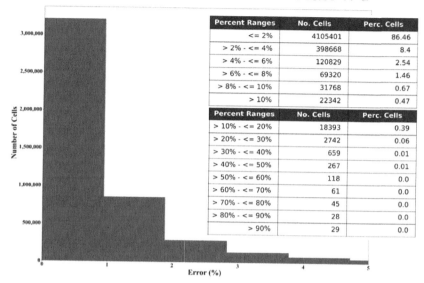

Percent Ranges	No. Cells	Perc. Cells
<= 2%	4105401	86.46
> 2% - <= 4%	398668	8.4
> 4% - <= 6%	120829	2.54
> 6% - <= 8%	69320	1.46
> 8% - <= 10%	31768	0.67
> 10%	22342	0.47

Percent Ranges	No. Cells	Perc. Cells
> 10% - <= 20%	18393	0.39
> 20% - <= 30%	2742	0.06
> 30% - <= 40%	659	0.01
> 40% - <= 50%	267	0.01
> 50% - <= 60%	118	0.0
> 60% - <= 70%	61	0.0
> 70% - <= 80%	45	0.0
> 80% - <= 90%	28	0.0
> 90%	29	0.0

Figure 5.116 Results of temperature distribution for blind validation CFD simulation run number 6 – Histogram.

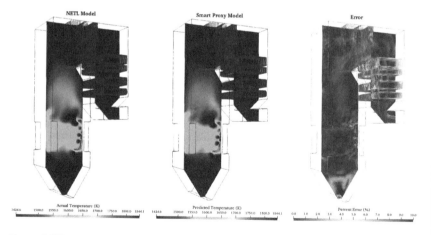

Figure 5.117 Temperature results for simulation run number 6 – Cross section number 1.

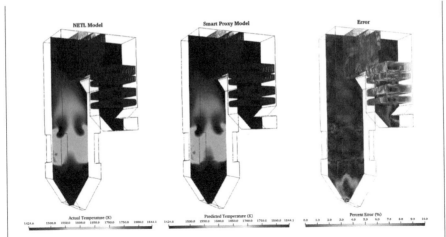

Figure 5.118 Temperature results for simulation run number 6 – Cross section number 2.

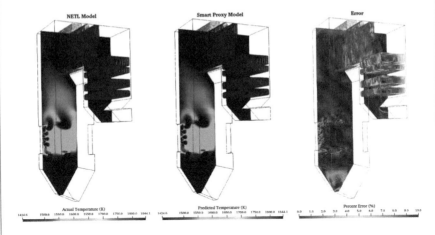

Figure 5.119 Temperature results for simulation run number 6 – Cross section number 3.

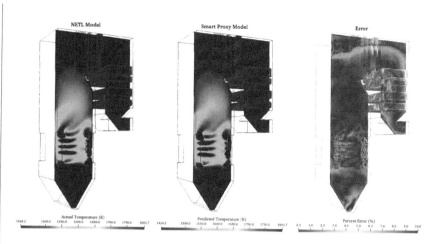

Figure 5.120 Temperature results for simulation run number 6 – Cross section number 4.

Then more results are presented to compare the Smart Proxy Model for temperature during the beginning of the project with the most recent enhanced version of the Smart Proxy Model for Tri-State boiler's CFD. These comparing temperature results are presented in Figures 5.121–5.130.

Smart Proxy Model: **Temperature**
Blind Validation – CFD Simulation Run #5

Base Model

Percent Ranges	No. Cells	Perc. Cells
<= 2%	2745839	57.83
> 2% - <= 4%	1282817	27.02
> 4% - <= 6%	437020	9.2
> 6% - <= 8%	138483	2.92
> 8% - <= 10%	61289	1.29
> 10%	82880	1.75

Percent Ranges	No. Cells	Perc. Cells
> 10% - <= 20%	70674	1.49
> 20% - <= 30%	7810	0.16
> 30% - <= 40%	2199	0.05
> 40% - <= 50%	1050	0.02
> 50% - <= 60%	514	0.01
> 60% - <= 70%	296	0.01
> 70% - <= 80%	161	0.0
> 80% - <= 90%	113	0.0
> 90%	63	0.0

Enhanced Model

Percent Ranges	No. Cells	Perc. Cells
<= 2%	4084306	86.02
> 2% - <= 4%	394600	8.31
> 4% - <= 6%	124188	2.62
> 6% - <= 8%	72290	1.52
> 8% - <= 10%	43611	0.92
> 10%	29333	0.62

Percent Ranges	No. Cells	Perc. Cells
> 10% - <= 20%	26097	0.55
> 20% - <= 30%	2335	0.05
> 30% - <= 40%	611	0.01
> 40% - <= 50%	181	0.0
> 50% - <= 60%	46	0.0
> 60% - <= 70%	33	0.0
> 70% - <= 80%	15	0.0
> 80% - <= 90%	6	0.0
> 90%	9	0.0

Figure 5.121 Comparing the results of based Smart Proxy Model versus enhanced Smart Proxy Model for temperature distribution of blind validation CFD simulation run number 5 – Histogram.

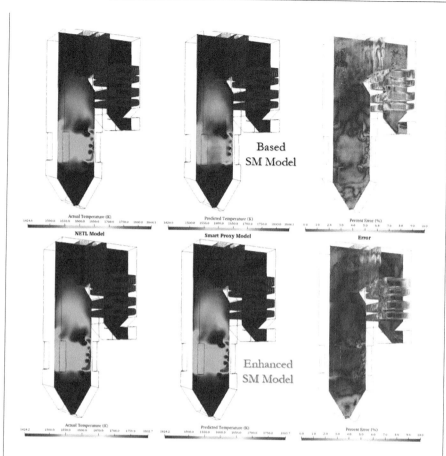

Figure 5.122 Comparing the temperature distribution between based Smart Proxy Model and enhanced Smart Proxy Model for blind validation CFD simulation run number 5 – Cross section number 1.

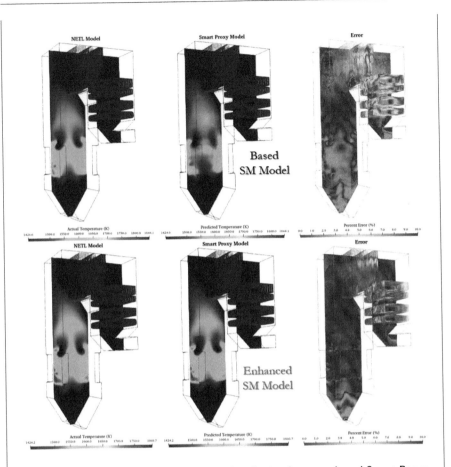

Figure 5.123 Comparing the temperature distribution between based Smart Proxy Model and enhanced Smart Proxy Model for blind validation CFD simulation run number 5 – Cross section number 2.

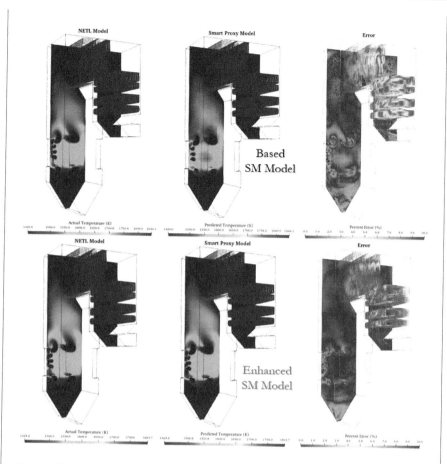

Figure 5.124 Comparing the temperature distribution between based Smart Proxy Model and enhanced Smart Proxy Model for blind validation CFD simulation run number 5 – Cross section number 3.

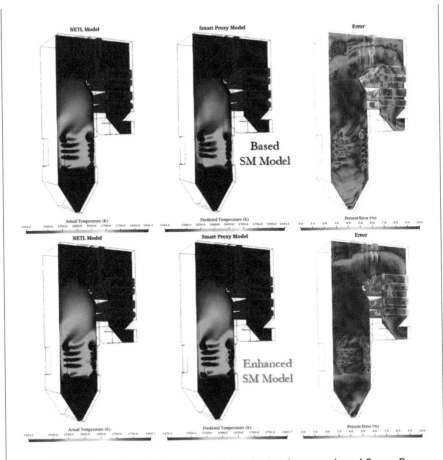

Figure 5.125 Comparing the temperature distribution between based Smart Proxy Model and enhanced Smart Proxy Model for blind validation CFD simulation run number 5 – Cross section number 4.

Smart Proxy Model: Temperature
Blind Validation – CFD Simulation Run #6

Base Model

Percent Ranges	No. Cells	Perc. Cells
<= 2%	2570572	54.14
> 2% - <= 4%	1373596	28.93
> 4% - <= 6%	427178	9.0
> 6% - <= 8%	162298	3.42
> 8% - <= 10%	71131	1.5
> 10%	143553	3.02

Percent Ranges	No. Cells	Perc. Cells
> 10% - <= 20%	124304	2.62
> 20% - <= 30%	13116	0.28
> 30% - <= 40%	3353	0.07
> 40% - <= 50%	1411	0.03
> 50% - <= 60%	709	0.01
> 60% - <= 70%	345	0.01
> 70% - <= 80%	184	0.0
> 80% - <= 90%	90	0.0
> 90%	41	0.0

Enhanced Model

Percent Ranges	No. Cells	Perc. Cells
<= 2%	4105401	86.46
> 2% - <= 4%	398668	8.4
> 4% - <= 6%	120829	2.54
> 6% - <= 8%	69320	1.46
> 8% - <= 10%	31768	0.67
> 10%	22342	0.47

Percent Ranges	No. Cells	Perc. Cells
> 10% - <= 20%	18393	0.39
> 20% - <= 30%	2742	0.06
> 30% - <= 40%	659	0.01
> 40% - <= 50%	267	0.01
> 50% - <= 60%	118	0.0
> 60% - <= 70%	61	0.0
> 70% - <= 80%	45	0.0
> 80% - <= 90%	28	0.0
> 90%	29	0.0

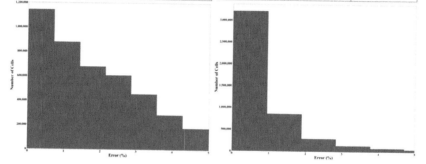

Figure 5.126 Comparing the results of based Smart Proxy Model versus enhanced Smart Proxy Model for temperature distribution of blind validation CFD simulation run number 6 – Histogram.

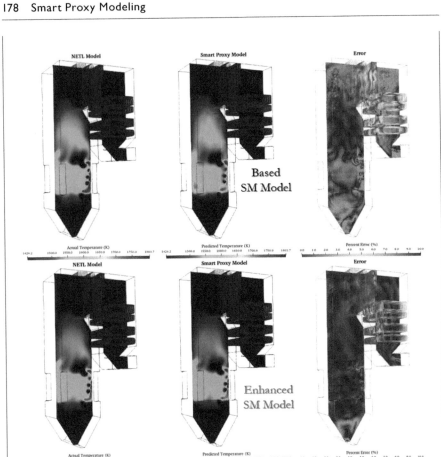

Figure 5.127 Comparing the temperature distribution between based Smart Proxy Model and enhanced Smart Proxy Model for blind validation CFD simulation run number 6 – Cross section number 1.

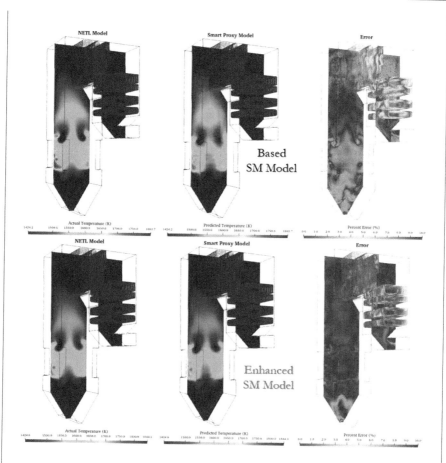

Figure 5.128 Comparing the temperature distribution between based Smart Proxy Model and enhanced Smart Proxy Model for blind validation CFD simulation run number 6 – Cross section number 2.

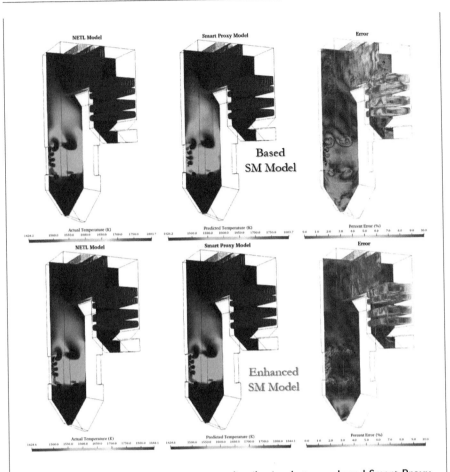

Figure 5.129 Comparing the temperature distribution between based Smart Proxy Model and enhanced Smart Proxy Model for blind validation CFD simulation run number 6 – Cross section number 3.

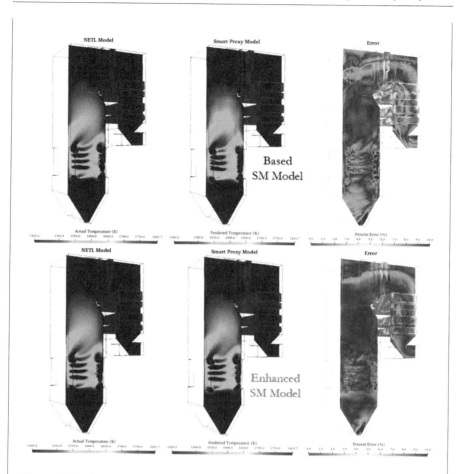

Figure 5.130 Comparing the temperature distribution between based Smart Proxy Model and enhanced Smart Proxy Model for blind validation CFD simulation run number 6 – Cross section number 4.

References

Al-Abbas, A.H.; Naser, J.; and Dodds, D., "CFD modelling of air-fired and oxy-fuel combustion in a large-scale furnace at Loy Yang A brown coal power station, Fuel," *Fuel*, vol. 102, pp. 646–665, 2012.

Al-Abbas, A.H.; Naser, J.; Dodds, D.; and Blicblau, A., "Numerical modelling of oxy-fuel combustion in a full-scale tangentially-fired pulverised coal boiler," *Procedia Engineering*, vol. 56, pp. 375–380, 2013.

Ansari, A., "Developing a Smart Proxy for Fluidized Bed Using Machine Learning," Graduate Theses, Dissertations, and Problem Reports. 5113, 2016.

Ansari, A.; Mohaghegh, S.; Shahnam, M.; Dietiker, J.F.; and Li, T., "Data Driven Smart Proxy for CFD Application of Big Data Analytics & Machine Learning in Computational Fluid Dynamics, Part Two: Model Building at the Cell Level," NETL-PUB-21634, NETL Technical Report Series; US Department of Energy, National Energy Technology Laboratory, Morgantown, WV, 2017.

ANSYS, Inc., ANSYS Fluent User's Guide, Release 16.1, Southpointe 2600 ANSYS Drive, Canonsburg, PA 15317, April 2015.

Aziz, K.; and Settari, A., *Petroleum Reservoir Simulation*, Applied Science Publishers Ltd., London, 1979.

Bases, G., "The Boiler Circulatory System: Beyond the Steam-Generating Boiler." *Insulation Outlook Magazine.* https://insulation.org/io/articles/the-boiler-circulatory-system-beyond-the-steam-generating-boiler/, 9 June 2017.

Bell, G.; Hey, T.; and Szalary, A., "Beyond the data deluge," 5919, *Science*, vol. 323, pp. 1297–1298, 2009.

Bhuiyan, A.; and Naser, J., "CFD modelling of co-firing of biomass with coal under oxy-fuel combustion in a large-scale power plant," *Fuel*, vol. 159, pp. 150–168, 2015.

Bruce, W.A., "An electrical device for analyzing oil reservoir behavior," *AIME Transactions*, vol. 157, p. 112, 1943.

Cardoso, M.A.; and Durlofsky, L.J., "Linearized reduced-order models for subsurface flow simulation," Department of Energy Resources Engineering, Stanford University, Stanford, CA 94305, USA, *Journal of Computational Physics*, vol. 229, pp. 681–700, 2010.

Chen, S.; He, B.; He, D.; Cao, Y.; Ding, G.; Liu, X.; Duan, Z.; Zhang, X.; Song, J.; and Li, X, "Numerical investigations on different tangential arrangements of burners for a 600 MW utility boiler," *Energy*, vol. 122, pp. 287–300, 2017.

Chen, H.; Klie, H.; and Wang, Q., "A Black-Box Stencil Interpolation Method to Accelerate Reservoir Simulations," SPE 163614, SPE Reservoir Simulation Symposium Held in The Woodlands, Texas, USA, 18–20 February 2013.

Choi, C., "Numerical investigation on the flow, combustion and NOx emission characteristics in a 500 MWe tangentially fired pulverized-coal boiler," *Fuel*, vol. 88, pp. 1720–1731, 2009.

Creosteanu, A.; Gavrila, G.; and Creosteanu L., "Comparison between an Analytical Method and Two Numerical Methods on a Given Electrostatic Potential Determination Problem," *15 International Symposium on Antenna Technology and Applied Electromagnetics, Toulouse, France*, DOI: 10.1109/ANTEM.2012.6262310 (IEEE INSPEC Accession Number: 12932685), 25–28 June 2012.

Edge, P.; Heggs, P.; Pourkashanian, M.; and Williams, A., "An integrated computational fluid dynamics–process model of natural circulation steam generation in a coal-fired power plant," *Computers and Chemical Engineering*, vol. 35, no. 12, pp. 2618–2631, 2011.

Fullmer, W.; and Hrenva, C., "Quantitative assessment of fine-grid kinetic theory-based predictions of mean-slip in unbounded fluidization," *AIChE Journal*, vol. 62, pp. 11–17, 2016. DOI: 10.1002/aic.

García, I.; Barragán, A.; and Román, M., "CFD Simulations as a Tool for Flow and Thermal Analysis in Boilers of Power Plants," 2012.

Ge, X.; Dong, J.; Fan, H.; Zhang, Z.; Shang, X.; Hu, X.; and Zhang, J., "Numerical investigation of oxy-fuel combustion in 700°C-ultra-supercritical boiler," *Fuel*, vol. 207, pp. 602–614, 2017.

Gu, H.; Zhu, H.; Cui, Y.; Si, F.; Xue, R.; Xi, H.; and Zhang, J., "Optimized scheme in coal-fired boiler combustion based on information entropy and modified K-prototypes algorithm," *Results in Physics*, vol. 9, pp. 1262–1274, Elsevier, 2018.

Gubba, S.; Ingham, D.; Larsen, K.; Ma, L.; and Pourkashanian, M., "Numerical modelling of the co-firing of pulverised coal and straw in a 300MWe tangentially fired boiler," *Fuel Processing Technology*, vol. 104, pp. 181–188, 2012.

Haghighat, S. Alireza; and Mohaghegh Shahab D., "Interpretation of real-time pressure measurements to detect CO2 leakage," *SPE Annual Technical Conference and Exhibition, OnePetro*, 2015.

Hashimoto, N.; Kurose, R.; Hwang, S.-M.; Tsuji, H.; and Shirai, H., "A Numerical Simulation of Pulverized Coal Combustion Employing a Tabulated-Devolatilization-Process Model (TDP Model)". Elsevier. Retrieved from www.elsevier.com/locate/combustflame, 2011.

He, J.; and Durlofsky, L., "Reduced-order modeling for compositional simulation by use of trajectory piecewise linearization," *SPE Journal*, vol. 19, no. 05, pp. 858–872, 2014.

Klie, H., "Unlocking Fast Reservoir Predictions via Non-Intrusive Reduced Order Models," SPE 163584, *SPE Reservoir Simulation Symposium Held in The Woodlands, Texas, USA*, 18–20 February 2013.

Modliński, N.; Madejski, P.; Janda, T.; Szczepanek, K.; and Kordylewski, W., "A validation of computational fluid dynamics temperature distribution prediction in a pulverized coal boiler with acoustic temperature measurement," *Energy*, vol. 92, pp. 77–86, 2015.

Mohaghegh, S.D., "Quantifying uncertainties associated with reservoir simulation studies using surrogate reservoir models," SPE 102492, *Proceedings, 2006 SPE Annual Conference & Exhibition. San Antonio, Texas*, 24–27 September 2006a.

Mohaghegh, S.D.; Hafez H.; Gaskari, S.; Haajizadeh, M.; and Kenawy, M., "Uncertainty analysis of a giant oil field in the Middle East using surrogate reservoir model," SPE 101474, Proceedings, 2006 Abu Dhabi International Petroleum Exhibition and Conference. Abu Dhabi, UAE, 5–8 November 2006b.

Mohaghegh, S.D.; Modavi, A.; Hafez, H.; Haajizadeh, M.; Kenawy, M.; and Guruswamy, S., "Development of surrogate reservoir models (SRM) for fast track analysis of complex reservoirs," SPE 99667, Proceedings, 2006 SPE Intelligent Energy Conference and Exhibition. Amsterdam, the Netherlands, 11–13 April 2006c.

Sanpasertparnich, T.; and Aroonwilas, A., "Simulation and optimization of coal-fired power plants," *Energy Procedia*, Elsevier, vol. 1, no. 1, pp. 3851–3858, February 2009.

Saripalli, R.; Wang, T.; and Day, B., "Simulation of combustion and thermal flow," Industrial Energy Technology Conference, New Orleans, 2005.

Shahnam, M.; Gel, A.; Dietiker, J.-F.; Subramaniyan, A.K.; and Musser, J., "The effect of grid resolution and reaction models in simulation of a fluidized bed gasifier through nonintrusive uncertainty quantification techniques," *ASME Journal of Verification, Validation and Uncertainty Quantification*, vol. 1, no. 4, pp. 041004, 2016. DOI: 10.1115/1.4035445.

Smith, T.F.; Shen, Z.F.; and Friedman, J.N., "Evaluation of coefficients for the weighted sum of gray gases model," *Heat Transfer*, vol. 104, no. 4, pp. 602–608, 1982.

Sun, W.; Zhong, W.; Yu, A.; Liu, L.; and Qian, Y., "Numerical investigation on the flow, combustion, and NOX emission characteristics in a 660 MWe tangential firing ultra-supercritical boiler," *Advances in Mechanical Engineering*, vol. 8, no. 2, p. 36, 2016.

Williams, G.J.J.; Mansfield, M.; MacDonald, D.G.; and Bush, M.D., "Top-down reservoir modelling," *Society of Petroleum Engineers*, 1 January 2004. DOI: 10.2118/89974-MS.

Wilson, K.C.; and Durlofsky, L.J., "Computational Optimization of Shale Resources Development Using Reduced Physics Surrogate Models," SPE 152946, SPE Western Regional Meeting, Bakersfield, CA. 19–23 March 2012.

Yin, C., "Refined weighted sum of gray gases model for air-fuel combustion and its impacts," *Energy Fuels*, vol. 27, pp. 6287–6294, 2013.

Zhou, H.; Yang, Y.; Liu, H.; and Hang, Q., "Numerical simulation of the combustion characteristics of a low NOx swirl burner: Influence of the primary air pipe," *Fuel*, vol. 130, pp. 168–176, 2014.

Index

Note: Locators in *italics* represent figures in the text.

9781032151144